CAMBRIDGE INTERNATIONAL AS AND A LEVEL

CHEMISTRY

David B

HODDER EDUCATION

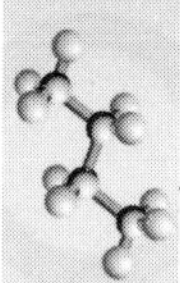

Hachette UK's policy is to use papers that are natural, renewable and recyclable products and made from wood grown in sustainable forests. The logging and manufacturing processes are expected to conform to the environmental regulations of the country of origin.

Orders: please contact Bookpoint Ltd, 130 Milton Park, Abingdon, Oxon OX14 4SB. tel: (44) 01235 827827; fax: (44) 01235 400401; email: education@bookpoint.co.uk. Lines are open 9.00–5.00, Monday to Saturday, with a 24-hour message answering service. Visit our website at www.hoddereducation.co.uk

First published in 2011 by
Hodder Education, a Hachette UK company
338 Euston Road
London NW1 3BH

Impression number 5 4 3 2 1
Year 2014 2013 2012 2011

Illustrations by Greenhill Wood Studios
Typeset in ITC Leawood 8.25pt by Greenhill Wood Studios
Printed by MPG Books, Bodmin

A catalogue record for this title is available from the British Library

ISBN 978 1 4441 1268 9

P01763

Contents

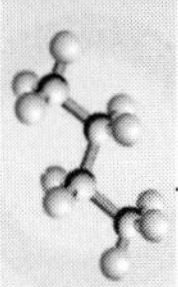

Experimental Skills & Investigations

■ ■ ■

Questions & Answers

■ ■ ■

Introduction

About this guide

This book is intended to help you to prepare for your University of Cambridge International AS and A level chemistry examinations. It is a revision guide, which you can use alongside your usual textbook as you work through your course, and also towards the end when you are revising for your examination.

The guide has four main sections:

- This **Introduction** contains an overview of the AS and A2 chemistry courses and how they are assessed, some advice on revision and advice on the question papers.
- The **Content Guidance** provides a summary of the facts and concepts that you need to know for the AS and A2 chemistry examinations.
- The **Experimental Skills** section explains the data-handling skills you will need to answer some of the questions in the written papers. It also explains the practical skills that you will need in order to do well in the practical examination.
- The **Questions and Answers** section contains practice examination papers for you to try. There is also a set of students' answers for each question, with comments from an examiner.

There are a number of ways to use this book. We suggest you start by reading through this Introduction, which will give you some suggestions about how you can improve your knowledge and skills in chemistry and about some good ways of revising. It also gives you pointers into how to do well in the examination. The Content Guidance will be especially useful when you are revising, as will the Questions and Answers.

The syllabus

It is a good idea to have your own copy of the University of Cambridge International Examinations (CIE) AS and A level chemistry syllabus. You can download it from

http://www.cie.org.uk/qualifications/academic/uppersec/alevel/

The **Syllabus Content** provides details of the chemical facts and concepts that you need to know, so keep a check on this as you work through your course. The Syllabus Content is divided into 25 sections, 1 to 11.3. Each section contains many learning outcomes. If you feel that you have not covered a particular learning outcome, or if you feel that you do not understand something, it is a good idea to work to correct this at an early stage. Don't wait until revision time!

Do look through all the other sections of the syllabus as well. There is a useful section on **Glossary of terms used in science papers** and another giving resources to help you in your study.

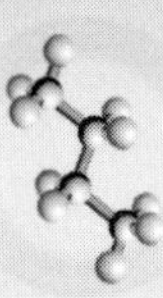

Syllabus content

The content of the A-level syllabus is divided into 25 sections:

Topic 1 Atoms, molecules and stoichiometry
Topic 2 Atomic structure
Topic 3 Chemical bonding
Topic 4 States of matter
Topic 5 Chemical energetics
Topic 6 Electrochemistry
Topic 7 Equilibria
Topic 8 Reaction kinetics
Topic 9.1 The periodic table: chemical periodicity
Topic 9.2 Group II
Topic 9.3 Group IV
Topic 9.4 Group VII
Topic 9.5 An introduction to the chemistry of transition elements
Topic 9.6 Nitrogen and sulfur
Topic 10.1 Introductory organic chemistry
Topic 10.2 Hydrocarbons
Topic 10.3 Halogen derivatives
Topic 10.4 Hydroxy compounds
Topic 10.5 Carbonyl compounds
Topic 10.6 Carboxylic acids and derivatives
Topic 10.7 Nitrogen compounds
Topic 10.8 Polymerisation
Topic 11.1 The chemistry of life
Topic 11.2 Applications of analytical chemistry
Topic 11.3 Design and materials

The main part of this book, the Content Guidance, summarises the facts and concepts covered by the learning outcomes in all of these 25 sections. Some of these sections deal only with AS material and some with just A2 material. Most chapters contain aspects of both AS and A2, and the A2 material is clearly indicated by a bar in the margin.

Assessment

The AS examination can be taken at the end of the first year of your course, or with the A2 examination papers at the end of the second year of your course.

What is assessed?

Both the AS and A2 examinations will test three Assessment Objectives. These are:

A: Knowledge with understanding

This involves your knowledge and understanding of the facts and concepts described in the learning outcomes in all of the 25 sections. Questions testing this Assessment Objective will make up 46% of the whole A-level examination.

B: Handling information and solving problems

This requires you to use your knowledge and understanding to answer questions involving unfamiliar contexts or data. The examiners ensure that questions testing this Assessment Objective cannot have been practised by candidates. You will have to *think* to answer these questions, not just remember! An important part of your preparation for the examination will be to gain confidence in answering this kind of question. Questions testing this Assessment Objective will make up 30% of the whole examination.

C: Experimental skills and investigations

This involves your ability to do practical work. The examiners set questions that require you to carry out experiments. It is most important that you take every opportunity to improve your practical skills as you work through your course. Your teacher should give you plenty of practice doing practical work in a laboratory. Questions testing this Assessment Objective will make up 24% of the whole A-level examination. This Assessment Objective is assessed in the AS practical paper (Paper 3) and in Paper 5 at A2.

Notice that more than half of the marks in the examination — 54% — are awarded for Assessment Objectives B and C. You need to work hard on developing these skills, as well as learning facts and concepts. There is guidance about this on pages 219–228 of this book.

The examination

The AS examination has three papers:

- Paper 1 — Multiple choice
- Paper 2 — Structured questions
- Paper 3 — Advanced practical skills

Paper 1 and Paper 2 test Assessment Objectives A and B. Paper 3 tests Assessment Objective C.

Paper 1 contains 40 multiple-choice questions. You have 1 hour to answer this paper. This works out at about one question per minute, with time left over to go back through some of the questions again.

Paper 2 contains structured questions. All the questions must be answered. You write your answers on the question paper. You have 1 hour 15 minutes to answer this paper.

Paper 3 is a practical examination. You will work in a laboratory. As for Paper 2, you write your answers on lines provided in the question paper. You have 2 hours to answer this paper.

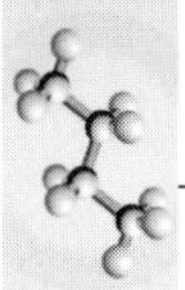

The A2 examination has two papers:

- Paper 4 Structured questions
- Paper 5 Planning, analysis and evaluation

Paper 4 tests Assessment Objectives A and B. Paper 5 tests Assessment Objective C.

Paper 4 has two sections and you have 2 hours to complete it. All the questions must be answered. You write your answers on the question paper.

Paper 5 contains a number of questions based on the practical skills, including planning, analysis and evaluation. As for Paper 4, you write your answers on lines provided in the question paper. Note that this is not a practical examination.

You can find copies of past papers at

http://www.cambridgestudents.org.uk/subjectpages/chemistry/asalchemistry/pastpapers/

Scientific language

Throughout your chemistry course, and especially in your examination, it is important to use clear and correct chemical language. Scientists take great care to use language precisely. If doctors or researchers do not use exactly the right word when communicating with someone, then what they say could easily be misinterpreted.

Chemistry has a huge number of specialist terms and symbols and it is important that you learn them and use them correctly.

However, the examiners are testing your knowledge and understanding of chemistry, not how well you can write in English. They will do their best to understand what you mean, even if some of your spelling and grammar is not correct. Nevertheless, there are some words that you really must spell correctly, because they could be confused with other chemical terms. These include:

- words that differ from one another by only one letter, for example ethane and ethene
- words with similar spellings but different meanings, for example homogeneous and homolytic; catalysis and catalase; sulfurous and sulfuric; phosphorus and phosphorous

In the Syllabus Content section of the syllabus, the words for which you need to know definitions are printed in *italic*. You will find definitions of most of these words in this text.

Revision

You can download a revision checklist at

http://www.cambridgestudents.org.uk/subjectpages/chemistry/asalchemistry/

This lists all of the learning outcomes, and you can tick them off or make notes about them as your revision progresses.

There are many different ways of revising, and what works well for you may not be as suitable for someone else. Have a look at the suggestions below and try some of them out.

- **Revise continuously**. Don't think that revision is something you do just before the exam. Life is much easier if you keep revision ticking along all through your chemistry course. Find 15 minutes a day to look back over work you did a few weeks ago, to keep it fresh in your mind. You will find this very helpful when you come to start your intensive revision.
- **Understand it**. Research shows that people learn things much more easily if the brain recognises that they are important and that they make sense. Before you try to learn a topic, make sure that you understand it. If you don't, ask a friend or a teacher, find a different textbook in which to read about it, or look it up on the internet. Work at it until you feel you have got it sorted and *then* try to learn it.
- **Make your revision active**. Just reading your notes or a textbook will not do any harm, but nor will it do much good either. Your brain only puts things into its long-term memory if it thinks they are important, so you need to convince it that they are. You can do this by making your brain *do* something with what you are trying to learn. So, if you are revising from a table comparing the reactions of alkanes and alkenes, try rewriting it as a paragraph of text, or converting it into two series of equations. You will learn much more by constructing your own list of bullet points, flow diagram or table than just trying to remember one that someone else has constructed.
- **Fair shares for all**. It is not a good idea to always start your revision in the same place. If you always start at the beginning of the course, then you will learn a great deal about atoms but not very much about organic chemistry or applications of chemistry. Make sure that each part of the syllabus gets its fair share of your attention and time.
- **Plan your time**. You may find it helpful to draw up a revision plan, setting out what you will revise and when. Even if you don't stick to it, it will give you a framework that you can refer to. If you get behind with it, you can rewrite the next parts of the plan to squeeze in the topics you have not yet covered.
- **Keep your concentration**. It is often said that it is best to revise in short periods, say 20 minutes or half an hour. This is true for many people who find it difficult to concentrate for longer than that. But there are others who find it better to settle down for a much longer period of time — even several hours — and really get into their work and stay concentrated without interruptions. Find out which works best for you. It may be different at different times of the day. Maybe you can concentrate well for only 30 minutes in the morning, but are able to get lost in your work for several hours in the evening.
- **Don't assume you know it**. The topics in which exam candidates are least likely to do well are often the ones that they have already learned something about at GCSE, IGCSE or O-level. This is probably because if you think you

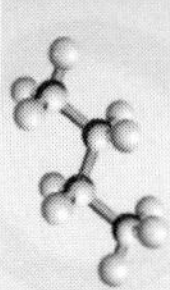

already know something then you give that a low priority when you are revising. It is important to remember that what you knew for your previous examinations is almost certainly not detailed enough for AS or A2.

The examination

Once you are in the examination room, you can stop worrying about whether or not you have done enough revision. Now you can concentrate on making the best use of the knowledge, understanding and skills that you have built up through your chemistry course.

Time

On average, you should allow about 1 minute for each mark on the examination paper.

In Paper 1, you will have to answer 40 multiple-choice questions in 1 hour. If you work to the 'one-mark-a-minute' rule, you should have plenty of time to look back over your answers and check any that you were not quite sure about. It is important to answer every question, even if you can only guess at the answer. Look carefully at the alternatives; you can probably eliminate one or two of the possible answers, which will increase your chances of your final guess being correct.

In Paper 2, you will have to answer 60 marks worth of short-answer questions in 75 minutes, so again there should be some time left over to check your answers at the end. In this paper it is probably worth spending a short time at the start of the examination to look through the whole paper. If you spot a question that you think may take you a little longer than others (for example, a question that has data to analyse), then you can make sure you allow plenty of time for this one.

In Paper 3, you will be working in a laboratory. You have 2 hours to answer 40 marks worth of questions. This is much more time per mark than in the other papers, but this is because you will have to do quite a lot of hands-on practical work before you obtain data to answer some of the questions. There will be two or three questions, and you should look at the breakdown of marks before deciding how long to spend on each question. Your teacher may split the class so that you have to move from one question to the other partway through the time allowed. It is easy to panic in a practical exam, but if you have done plenty of practical work throughout your course this will help you a lot. Do read through the whole question before you start, and do take time to set up your apparatus correctly and to collect your results carefully and methodically.

Paper 4 consists of two sections with a total of 100 marks. With only 2 hours to complete the paper you need to think clearly and carefully about your answers. The questions in Section A are based on the A2 syllabus, but may also include material from AS. The questions in Section B are more of the problem-solving type and may require you to use chemical knowledge from anywhere in the syllabus in new situations. These questions take a little longer than 1 minute per mark.

Paper 5 consists of a variable number of questions (usually two or three) that are based on the practical skills of planning, analysis and evaluation. The paper is 1 hour 15 minutes long, which seems quite generous for the 30 marks available. However, the questions will require some thought before you answer.

Read the question carefully

That sounds obvious, but candidates lose many marks by not reading questions carefully.

- There is often vital information at the start of the question that you'll need in order to answer the questions themselves. Don't just jump straight to the first place where there are answer lines and start writing. Begin by reading from the beginning of the question. Examiners are usually very careful not to give you unnecessary information, so if it is there then it is probably needed. You may like to use a highlighter to pick out any particularly important pieces of information at the start of the question.
- Look carefully at the command words at the start of each question, and make sure that you do what they say. For example if you are asked to *explain* something and you only *describe* it, you will not get many marks — indeed, you may not get any marks at all, even if your description is a very good one. You can find the command words and their meanings towards the end of the syllabus.
- Do watch out for parts of questions that don't have answer lines. For example, you may be asked to label something on a diagram, or to draw a line on a graph, or to write a number in a table. Many candidates miss out these questions and lose a significant number of marks.

Depth and length of answer

The examiners give you two useful guidelines about how much you need to write.

- **The number of marks.** The more marks, the more information you need to give in your answer. If there are 2 marks, you will need to give at least two pieces of correct and relevant information in your answer in order to get full marks. If there are 5 marks, you will need to write much more. But don't just write for the sake of it — make sure that what you write *answers the question.* And don't just keep writing the same thing several times over in different words.
- **The number of lines.** This isn't such a useful guideline as the number of marks, but it can still help you to know how much to write. If you find that your answer will not fit on the lines, then you have probably not focused sharply enough on the question. The best answers are short, precise, use correct chemical terms and don't repeat information already given.

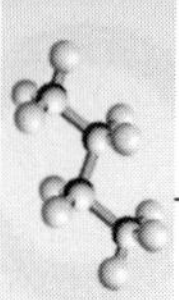

Writing, spelling and grammar

The examiners are testing your knowledge and understanding of chemistry, not your ability to write English. However, if they cannot understand what you have written, they cannot give you any marks. It is your responsibility to communicate clearly. Don't scribble so fast that the examiner cannot read what you have written. Every year, candidates lose marks because the examiner could not read their writing.

Like spelling, grammar is not taken into consideration when marking your answers — so long as the examiner can understand what you are trying to say. One common difficulty is if you use the word 'it' in your answer, and the examiner is not sure what you are referring to. For example, imagine a candidate writes 'Calcium metal dissolves easily in hydrochloric acid. It is very reactive.' Does the candidate mean that the calcium is very reactive, or that the hydrochloric acid is very reactive? If the examiner cannot be sure, you may not be given the benefit of the doubt.

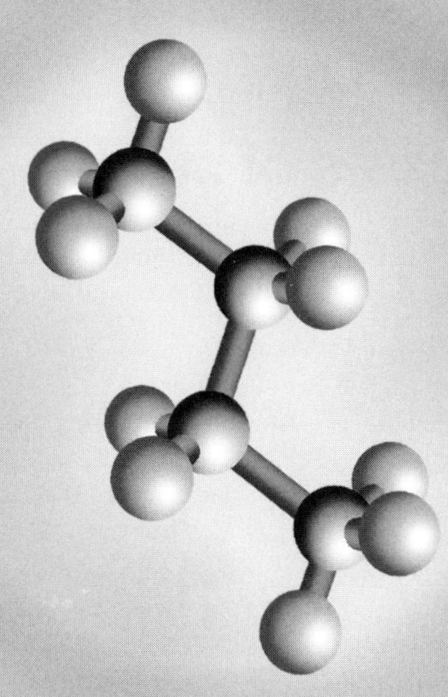

Content Guidance

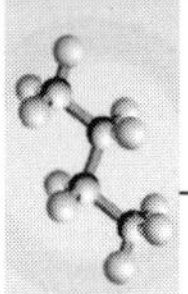

1 Atoms, molecules and stoichiometry

Relative masses of atoms

There are more than 100 chemical elements, and each element is made up of atoms. The atoms of different elements differ in size, and hence have different masses.

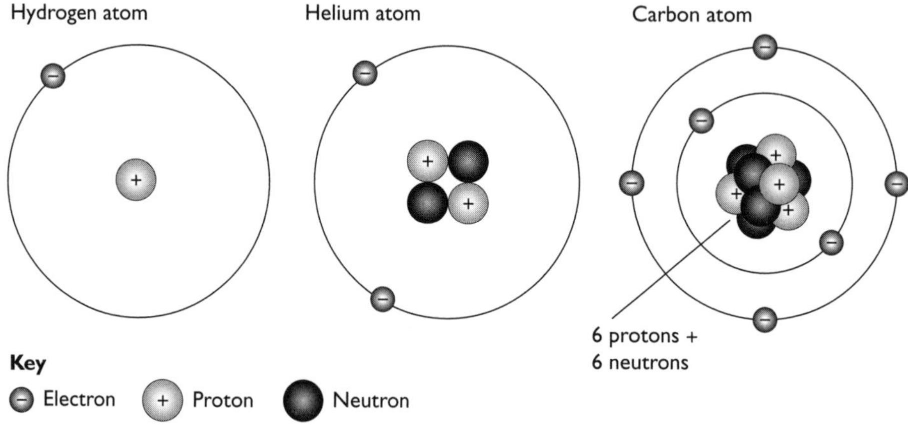

Figure 1.1 Atoms of hydrogen, helium and carbon

You can also see that the atoms are made up of different sorts and numbers of particles. There is more about this in the section on atomic structure. For now you should be able to identify:

- two types of particle in the **nucleus**, which is in the middle of the atom. The two particles in the nucleus are **protons** and **neutrons**. Both have the same mass but a proton has a single positive charge and a neutron has no charge.
- another type of particle that circles the nucleus. These particles are called **electrons**. An electron has almost no mass, but carries a single negative charge (Table 1.1).

Table 1.1

Particle	Relative mass	Relative charge
Proton	1	+1
Neutron	1	0
Electron	0	–1

Try this yourself

(1) Use a copy of the periodic table to work out which atoms are represented by the particles described in the table below. The final entry needs some careful thought. Can you work out what is going on here? (Answers are on p. 269.)

	Protons	Neutrons	Electrons	Identity of species
(a)	11	12	11	$^{23}_{11}Na$
(b)	9	10	9	
(c)	16	16	16	
(d)	24	28	24	
(e)	19	20	18	

For AS you need to be able to distinguish between terms that relate to the masses of elements and compounds.

Relative atomic mass, A_r, is defined as the mass of one atom of an element relative to 1⁄12 of the mass of an atom of carbon-12, which has a mass of 12.00 atomic mass units.

Relative isotopic mass is like relative atomic mass in that it deals with atoms. The difference is that we are dealing with *different forms* of the same element. Isotopes have the same number of protons, but different numbers of neutrons. Hence, isotopes of an element have different masses.

Relative molecular mass, M_r, is defined as the mass of one molecule of an element or compound relative to 1⁄12 of the mass of an atom of carbon-12, which has a mass of 12.00 atomic mass units.

Relative formula mass is used for substances that do not contain molecules, such as sodium chloride, Na*Cl*, and is the sum of all the relative atomic masses of the atoms present in the formula of the substance.

It is important to remember that since these are all *relative* masses, they have no units.

The mole

Individual atoms cannot be picked up or weighed, so we need to find a way to compare atomic masses. One way is to find the mass of the same number of atoms of different types. Even so, the mass of an atom is so small that we need a huge number of atoms of each element to weigh. This number is called the **Avogadro constant**. It is equal to 6.02×10^{23} atoms and is also referred to as **one mole**. You may wonder why such a strange number is used. It is the number of atoms of a substance that make up the relative atomic mass, A_r, in grams. The mass is measured relative to one-twelfth of the mass of a carbon atom, ^{12}C.

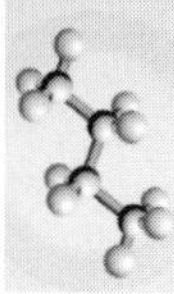

Mole calculations

You should be able to work out how many moles a given mass of an element or compound represents. In order to do that you need to know the relative atomic mass, A_r, of the element (or elements) present. You can get this information from the periodic table.

The abbreviation for mole is 'mol'.

Try this yourself

(2) How many moles do the following masses of *atoms* represent?
- **(a)** 6 g of carbon, C
- **(b)** 24 g of oxygen, O
- **(c)** 14 g of iron, Fe

(3) How many grams of substance are in the following?
- **(a)** 0.2 mol of neon, Ne
- **(b)** 0.5 mol of silicon, Si
- **(c)** 1.75 mol of helium, He
- **(d)** 0.25 mol of carbon dioxide, CO_2

Mass spectra

Another way of determining the atomic mass of an element is to use a **mass spectrometer**. You do not need to know how the instrument works, only that it produces positive ions of atoms or fragments of molecules and separates them according to their masses. Molecular fragmentation is covered later.

On placing a sample of an element in a mass spectrometer, atoms of the element become positively charged and separated according to their masses. Many elements are made up of atoms with the same number of protons but different numbers of neutrons. This means that they have different masses. The data can be used to calculate the average atomic mass of the sample. Figure 1.2 shows the mass spectrum of a sample of the element magnesium.

The average atomic mass of the sample of magnesium is made up of the contribution each isotope makes, i.e.

$$A_r = (24 \times 0.79) + (25 \times 0.10) + (26 \times 0.11) = 24.32$$

Remember that samples may not always contain just one isotope, or even the same mix of isotopes.

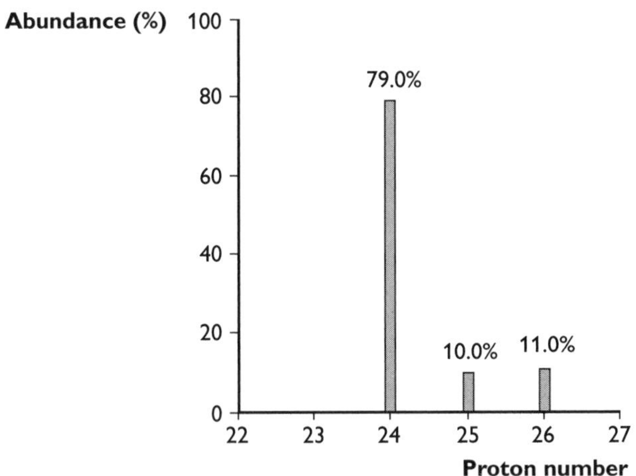

Figure 1.2 Mass spectrum of magnesium

Empirical and molecular formulae

The **empirical formula** of a compound is its simplest formula. It shows the ratio of the number of atoms of different elements in a compound. You need to know how to use the composition by mass of a compound to find its empirical formula:

- divide the mass (or percentage mass) of each element by its A_r
- use the data to calculate the simplest whole number ratio of atoms

Example

A chloride of iron contains 34.5% by mass of iron. Determine the empirical formula of the chloride.

Answer

Element	% by mass (*m*)	A_r	m/A_r	Moles	Ratio of moles of the elements
Fe	34.5	56	34.5/56	0.616	1
Cl	65.5	35.5	65.5/35.5	1.85	3

Thus the empirical formula of this chloride is $FeCl_3$.

Try this yourself

(4) Find the empirical formulae for the following compounds:

(a) Compound A — composition by mass: 84.2% rubidium, 15.8% oxygen

(b) Compound B — composition by mass: 39.1% carbon, 52.2% oxygen, 8.70% hydrogen

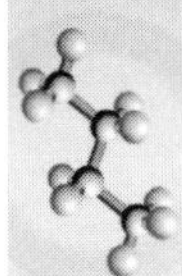

By contrast, the **molecular formula** of a compound shows the *actual number* of atoms of each element present in the compound. The molecular formula is always a multiple of the empirical formula.

Example

A compound has the empirical formula CH_2O, and a molar mass of 60. What is its molecular formula?

Answer

Using this information, we can see that the formula mass of the compound is

$(1 \times 12) + (2 \times 1) + (1 \times 16) = 30$

Since the molar mass is 60, the molecular formula must be twice the empirical formula, i.e. $C_2H_4O_2$.

Writing and balancing equations

Chemical equations are a shorthand way of describing chemical reactions. Using the symbols for elements from the periodic table ensures that they are understood internationally. Whenever you write a chemical equation there are simple rules to follow:

- check the formula of each compound in the equation
- check that the overall equation balances
- try to visualise what is happening in the reaction. This will help you choose the correct **state symbol** — the state symbols are (s) for solid, (l) for liquid, (g) for gas and (aq) for an aqueous solution

Suppose you want to write a chemical equation for the reaction between magnesium and dilute sulfuric acid. You can probably write a word equation for this from earlier in your studies of chemistry:

magnesium + sulfuric acid → magnesium sulfate + hydrogen

In symbols this becomes:

$Mg + H_2SO_4 \rightarrow MgSO_4 + H_2$

Counting up the number of each type of atom on each side of the arrow shows that they are equal. The equation is balanced.

You can include more detail if you think of the states of the reactants and products and add the state symbols:

$Mg(s) + H_2SO_4(aq) \rightarrow MgSO_4(aq) + H_2(g)$

You might also remember that dilute sulfuric acid is a mixture of H^+ and SO_4^{2-} ions. So we can write an ionic equation showing just the changes in species or chemical forms:

$$Mg(s) + 2H^+(aq) \rightarrow Mg^{2+}(aq) + H_2(g)$$

A more complicated reaction is the reaction between sodium carbonate and hydrochloric acid. You will have seen the mixture fizz in the laboratory:

sodium carbonate + hydrochloric acid → sodium chloride + carbon dioxide

In symbols this becomes:

$$Na_2CO_3 + HCl \rightarrow NaCl + CO_2$$

Counting the atoms on each side of the arrow, shows that there are 'spare' atoms of sodium, oxygen and hydrogen on the left-hand side and no hydrogen on the right-hand side. We can take care of the sodium by doubling the amount of sodium chloride formed, but what about the hydrogen and oxygen? Since water is a simple compound of hydrogen and oxygen, let's see what happens if water is added to the right-hand side:

$$Na_2CO_3 + 2HCl \rightarrow 2NaCl + CO_2 + H_2O$$

Doubling the amount of HCl and $NaCl$ now makes the equation balance. Adding the state symbols gives:

$$Na_2CO_3(s) + 2HCl(aq) \rightarrow 2NaCl(aq) + CO_2(g) + H_2O(l)$$

Notice that water is a liquid, not aqueous.

The ionic equation for this reaction is:

$$(Na^+)_2\ CO_3^{2-}(s) + 2H^+(aq) \rightarrow 2Na^+(aq) + CO_2(g) + H_2O(l)$$

Calculations using equations and the mole

Now that you understand moles and how to write balanced chemical equations, we can use these two ideas to calculate the quantities of substances reacting together and the amounts of products formed in reactions.

There are three main types of calculation you might be expected to perform in A-level chemistry:

- reacting masses (from formulae and equations)
- volumes of gases reacting or produced
- volumes and concentrations of solutions of chemicals reacting

In each of these you will need to use balanced chemical equations and the mole concept for quantities of chemical compounds.

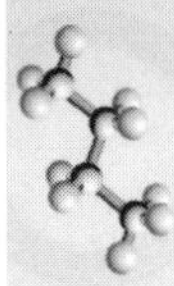

Calculations involving reacting masses

Suppose copper(II) carbonate is heated. What mass of copper(II) oxide would be formed starting from 5.0 g of the carbonate?

Let's break the calculation down into simple stages.

1 Write the equation for the reaction:

$CuCO_3 \rightarrow CuO + CO_2$

2 Now work out the relative molecular masses of each of the substances involved in the question:

$CuCO_3 \rightarrow CuO$

$63.5 + 12 + (3 \times 16) \rightarrow 63.5 + 16$

$123.5\,g \rightarrow 79.5\,g$

3 Finally, calculate the mass of CuO formed from 5.0 g of $CuCO_3$:

$5.0\,g \rightarrow \frac{5 \times 79.5}{123.5}\,g = 3.2\,g$

mass of CuO = 3.2 g

Tip If you try this calculation your calculator will show a lot more decimal places than are given in the answer of 3.2 g. The answer is given as 3.2 g because we use the number of significant figures equal to the smallest number of these in the data. Since the starting mass of copper(II) carbonate and the molar mass of carbon dioxide are quoted to two significant figures, we are not justified in giving an answer to more than two significant figures. This idea is important in scientific calculations. You will also come across its use in practical work involving calculations.

Try this yourself

Try the following calculations using the idea of reacting masses (remember to use the correct number of significant figures).

(5) What mass of carbon dioxide is lost when 2.5 g of magnesium carbonate is decomposed by heating?

(6) What mass of potassium chloride is formed when 2.8 g of potassium hydroxide is completely neutralised by hydrochloric acid?

(7) What is the increase in mass when 6.4 g of calcium is completely burned in oxygen?

The questions above are relatively straightforward. However, you might be asked to use mass data to determine the formula of a compound. The next example shows you how to do this.

Example

When heated in an inert solvent, tin metal reacts with iodine to form a single orange-red solid compound. In an experiment, a student used 5.00 g of tin metal in this reaction. After filtering and drying, the mass of crystals of the orange compound was 26.3 g. Using the data, work out the formula of the orange compound.

Answer

First you need to calculate how much iodine was used in the reaction. Do this by subtracting the mass of tin from the final mass of the compound:

mass of iodine used = 26.3 g – 5.00 g = 21.3 g

Next, convert the masses of tin and iodine into the number of moles of each. Do this by dividing each mass by the relevant atomic mass:

moles of tin = $\frac{5.00}{119}$ = 0.0420 mol

moles of iodine = $\frac{21.3}{127}$ = 0.168 mol

As you can see, the ratio of the number of moles shows that there are four times as many moles of iodine as there are tin in the compound. Therefore, the formula of the orange-red crystals is SnI_4.

Calculations involving volumes of gases

Not all chemical reactions involve solids. For those reactions in which gases are involved it is more convenient to measure volumes than masses. We need a way of linking the volume of a gas to the number of particles it contains — in other words a way to convert volume to moles. In the early nineteenth century, Avogadro stated that equal volumes of gases at the same temperature and pressure contain equal numbers of molecules. We now know that one mole of a gas occupies 24 dm^3 at room temperature (25 °C) and a pressure of 101 kPa (1 atm), or 22.4 dm^3 at standard temperature (273 K) and the same pressure (s.t.p.).

This means that if we measure the volume of gas in dm^3 in a reaction at room temperature and pressure, it can be converted directly to the number of moles present simply by dividing by 24.

The easiest way to see how this method works is to look at an example. Take the reaction between hydrogen and chlorine to form hydrogen chloride:

$$H_2(g) + Cl_2(g) \rightarrow 2HCl(g)$$

It would not be easy to measure the reacting masses of the two gases. We could, however, measure their volumes. When this is done, we find that there is no overall change in volume during the reaction. This is because there are two moles of gas on the left-hand side of the equation and two new moles of gas on the right-hand side.

Some reactions produce gases as well as liquids, and in others gases react with liquids to form solids, and so on. In these cases, we can use the above method combined with the method of the first calculation.

For example, 2.0 g of magnesium dissolves in an excess of dilute hydrochloric acid to produce hydrogen:

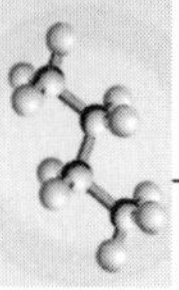

$Mg(s) + 2HCl(aq) \rightarrow MgCl_2(aq) + H_2(g)$

The equation shows that for every mole of magnesium used, 1 mole of hydrogen gas is formed.

Since 2.0 g of magnesium is 2.0/24.3 moles, this means that 2.0/24.3 moles of hydrogen gas should be formed.

Each mole of hydrogen occupies 24 dm^3 at room temperature and pressure:

volume of hydrogen produced $= \frac{2.0}{24.3} \times 24\,\text{dm}^3 = 1.98\,\text{dm}^3$

Try this yourself

Try the following calculations involving volumes of gas(es).

(8) 25 cm^3 of the gas propane, C_3H_8, is burnt in an excess of oxygen to form carbon dioxide and water. What volume of oxygen reacts, and what volume of carbon dioxide is formed at room temperature and pressure? (You may assume that the water formed is liquid and of negligible volume).

(9) A sample of lead(IV) oxide was heated in a test tube and the oxygen gas released was collected. What mass of the oxide would be needed to produce 80 cm^3 of oxygen at room temperature and pressure?
$2PbO_2(s) \rightarrow 2PbO(s) + O_2(g)$

(10) Carbon dioxide was bubbled into limewater (a solution of calcium hydroxide) and the solid calcium carbonate precipitated was filtered off, dried and weighed. If 0.50 g of calcium carbonate were formed what volume of carbon dioxide, at room temperature and pressure, was passed into the solution?
$Ca(OH)_2(aq) + CO_2(g) \rightarrow CaCO_3(s) + H_2O(l)$

Calculations involving volumes and concentrations of solutions

These types of calculation are particularly important since they often occur in the AS practical paper (see also the section on practical work). The basic principles of the calculations are the same as those covered already, the only complication being that the reactants are in solution. This means that instead of dealing with masses, we are dealing with volumes of solution of known molarity.

Another way of dealing with this is to see how many **moles** of substance are dissolved in 1 dm^3 of solution. This is known as the **molar concentration**. Do not confuse this with **concentration**, which is the **mass** of substance dissolved in 1 dm^3.

Think about a 0.1 mol dm^{-3} solution of sodium hydroxide. The mass of 1 mole of sodium hydroxide is (23 + 16 + 1) or 40 g. So a 0.1 mol dm^{-3} solution contains 0.1 mol (40 × 0.1 = 4.0 g) per dm^3 of solution.

If you know the molar concentration of a solution and the volume that reacts with a known volume of a solution containing another reactant, you can calculate the molar concentration of the second solution using the equation for the reaction.

Example

In a titration between dilute sulfuric acid and 0.1 molar sodium hydroxide, $21.70\,cm^3$ of the sodium hydroxide was needed to neutralise $25.00\,cm^3$ of the dilute sulfuric acid. Knowing the equation for the reaction, we can calculate the molar concentration of the acid in $mol\,dm^{-3}$:

$$H_2SO_4(aq) + 2NaOH(aq) \rightarrow Na_2SO_4(aq) + 2H_2O(l)$$

Answer

From the equation you can see that 1 mole of sulfuric acid requires 2 moles of sodium hydroxide for complete reaction.

number of moles of sodium hydroxide used = $\frac{21.70 \times 0.1}{1000}$

This would neutralise $\frac{21.70 \times 0.1}{1000 \times 2}$ moles of sulfuric acid

This number of moles is contained in $25.00\,cm^3$ sulfuric acid.

To get the number of moles in $1\,dm^3$, multiply this number by $\frac{1000}{25.00}$

This gives $\frac{21.70 \times 0.1 \times 1000}{1000 \times 2 \times 25.00} = 0.0434\,mol\,dm^{-3}$

Try this yourself

The following calculations involving volumes and concentrations of solutions will give you practice at this important area of the syllabus.

(11) In a titration, $27.60\,cm^3$ of $0.100\,mol\,dm^{-3}$ hydrochloric acid neutralised $25.00\,cm^3$ of potassium hydroxide solution. Calculate the molar concentration of the potassium hydroxide solution in $mol\,dm^{-3}$ and its concentration in $g\,dm^{-3}$.

(12) A $0.2\,mol\,dm^{-3}$ solution of nitric acid was added to an aqueous solution of sodium carbonate. $37.50\,cm^3$ of the acid were required to react completely with $25.00\,cm^3$ of the carbonate. Calculate the molar concentration of the carbonate in $mol\,dm^{-3}$.

2 Atomic structure

Subatomic particles and their properties

In Chapter 1 you saw that atoms are made up of three different types of particle — protons, neutrons and electrons. You should remember that only the protons and neutrons have significant mass, and that the proton carries a single positive charge while the electron carries a single negative charge. You also need to remember that protons and neutrons are found in the nucleus of the atom and that electrons surround the nucleus.

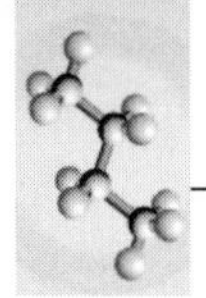

Look at the numbers of subatomic particles in the three particles shown in Table 2.1. What is the major difference between these three species?

Table 2.1

Particle	Number of protons	Number of neutrons	Number of electrons
A	11	12	10
B	11	12	11
C	11	12	12

The difference is in the number of electrons each particle possesses, and hence the overall charge on the species. Since **A** has one more proton than electron, it has a single positive charge. In **B** the numbers of protons and electrons are the same so it is uncharged (neutral). In **C** there is one more electron than proton, so it has a single negative charge. Notice that since all species have the same number of protons (proton number), they are all forms of the same element, in this case sodium. The two charged species are called ions:

- a positive ion is called a **cation**
- a negative ion is called an **anion**

You might be surprised to see sodium as an anion, Na^-, but it is theoretically possible (though very unlikely!).

Table 2.2 shows another way in which the numbers of subatomic particles can vary.

Table 2.2

Particle	Number of protons	Number of neutrons	Number of electrons
D	12	12	12
E	12	13	12
F	12	14	12

In this case, it is the number of neutrons that changes while the element stays the same. These forms of an element are called **isotopes**. In Table 2.2, the three are all isotopes of magnesium.

The standard way of writing these particles in 'shorthand' form is $^{M}_{P}X^{Y}$.

In this form the element symbol is X, *M* is the nucleon or mass number (the number of protons plus neutrons in the nucleus), *P* is the proton or atomic number (the number of protons in the nucleus) and *Y* is the charge (if any) on the particle.

Try this yourself

(13) Write out structures of the six species **A–F** described above using the form $^{M}_{P}X^{Y}$.

Arrangement of electrons in atoms

As the number of protons in the nucleus increases, the mass of an atom of the element increases. After hydrogen, this increase in mass is also due to the neutrons in the nucleus (see Table 2.3).

Table 2.3

Element	Protons	Neutrons	Relative atomic mass
H	1	0	1
He	2	2	4
Li	3	4	7
Be	4	5	9
B	5	6	11
C	6	6	12

The addition of electrons to form new atoms is not quite so straightforward because they go into different **orbitals** — regions in space that can hold a certain number of electrons, and which have different shapes. The electrons also exist in different **energy levels** (sometimes called shells) depending on how close to, or far away from, the nucleus they are.

The number of protons in the nucleus determines what the element is. However, it is the arrangement of electrons that determines the chemistry of an element and how it forms bonds with other elements. So, for example, metal atoms tend to lose electrons forming positive ions, and non-metal atoms tend to accept electrons forming negative ions.

As the number of protons increases, the electron energy levels fill up in the following sequence: 1s, 2s, 2p, 3s, 3p, 4s, 3d, 4p... (see Table 2.4). This sequence can be followed in the periodic table.

Table 2.4

Element	Electronic configuration
Hydrogen	$1s^1$
Helium	$1s^2$
Lithium	$1s^2\ 2s^1$
Beryllium	$1s^2\ 2s^2$
Boron	$1s^2\ 2s^2\ 2p^1$
Carbon	$1s^2\ 2s^2\ 2p^2$
Nitrogen	$1s^2\ 2s^2\ 2p^3$
Oxygen	$1s^2\ 2s^2\ 2p^4$
Fluorine	$1s^2\ 2s^2\ 2p^5$
Neon	$1s^2\ 2s^2\ 2p^6$

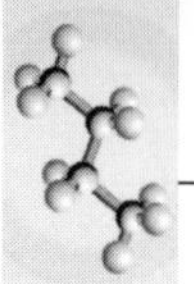

As more electrons are added, they go into orbitals of increasing energy:

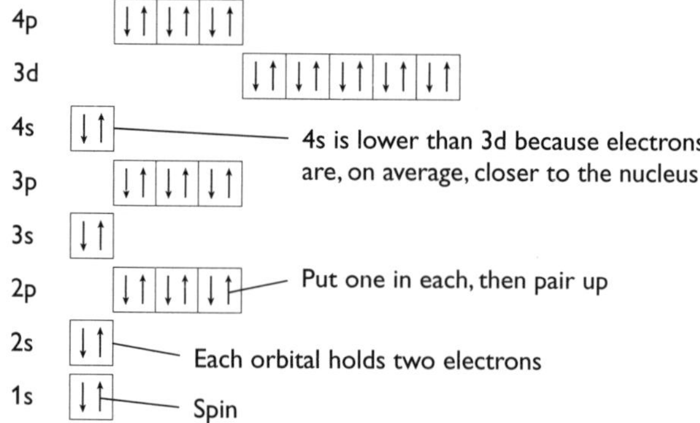

Figure 2.1 Sequence of filling orbitals with electrons

Figure 2.1 illustrates some key points in the arrangement of electrons in atoms. These are things you should remember:

- The electrons are arranged in **energy levels** (or shells) from level 1, closest to the nucleus. On moving outwards from the nucleus, the shells gradually increase in energy.
- Most energy levels (except the first) contain sub-levels (or sub-shells) denoted by s, p and d.
- Different sub-levels contain different numbers of orbitals, with each orbital holding a maximum of two electrons.
- When filling up the energy levels in an atom, electrons go into the lowest energy level first.
- In sub-levels containing more than one orbital, each of the orbitals is filled singly before any are doubly-filled.
- If you study Figure 2.1 there is one strange entry — the 4s-orbital has lower energy than the 3d-orbital.

In an examination, you may be asked to deduce the electron configuration of an atom (or ion) given its proton number (and any charge). The following examples show how to do this.

Example 1

An atom, **X**, has a proton number of 16. Deduce the electron configuration of this atom.

Answer

The proton number is 16, so the atom must also contain 16 electrons. Referring back to Figure 2.1, we can count upwards until 16 electrons have been used. This means the 1s-orbital, the 2s- and 2p-orbitals and the 3s-orbital are filled, using up 12 of the 16 electrons. The remaining four electrons must go into the 3p-orbital, giving the electron configuration $1s^2$, $2s^2$, $2p^6$, $3s^2$, $3p^4$.

Example 2

X forms an ion, $\mathbf{X}^{2-}$. What is the electron configuration of the ion?

Answer

The ion contains an extra two electrons compared with the atom. This means that it contains a total of (16 + 2) or 18 electrons. Looking at Figure 2.1 you can see that these extra two electrons will fit into the remainder of the 3p-orbital giving an electronic configuration of $1s^2$, $2s^2$, $2p^6$, $3s^2$, $3p^6$ for the ion $\mathbf{X}^{2-}$.

The different orbitals have different shapes. Cross-sections of these orbitals are shown in Figure 2.2.

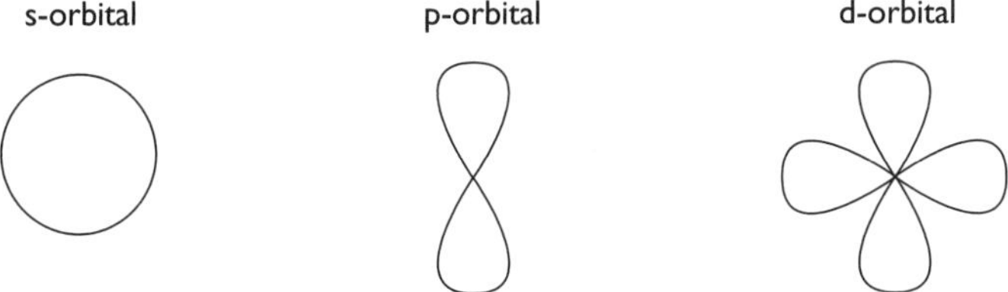

Figure 2.2 Cross sections of s-, p- and d-orbitals

The location of electrons in the different types of orbital can affect the shapes of molecules.

Ionisation energies

The first ionisation energy of an atom has a precise definition that you need to remember.

It is the energy required to convert 1 mole of gaseous atoms of an element into 1 mole of gaseous cations, with each atom losing one electron. This can be represented as follows:

$$M(g) \rightarrow M^+(g) + e^-$$

As you might expect, there is a change in the first ionisation energy as the number of protons in the nucleus increases. This leads to a '2-3-3' pattern for periods 2 and 3, as shown in Figure 2.3.

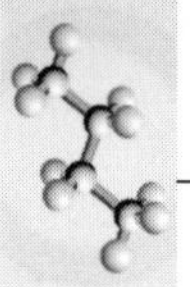

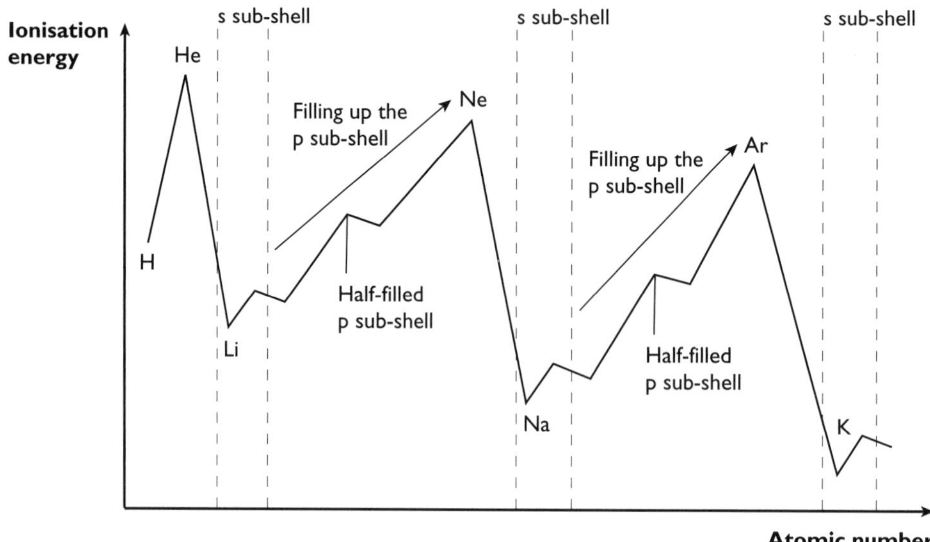

Figure 2.3 Relationship between first ionisation energy and atomic number

The graphs shown in some textbooks look complicated. For the examination you need to know the principles of the change. In an examination you might be asked to explain:

- the general increase in first ionisation energy across a period — proton number/nuclear charge increases across the period; shielding by other electrons is similar, hence there is a greater attraction for the electrons
- the drop between groups 2 and 3 and/or groups 5 and 6 — for groups 2 and 3 the electron is removed from a p-orbital, which is higher in energy than an electron in an s-orbital, so it is easier to remove. For groups 5 and 6 there are paired electrons in one of the p-orbitals. This causes repulsion, which makes it easier to remove one of these electrons.
- the big drop at the end of the period — an extra electron shell has been completed, which results in more shielding. Hence there is less attraction for the outer electrons.

Successive ionisation energies

Successive ionisation energies refer to the removal of second and subsequent electrons, for example:

Second ionisation energy $M^{+}(g) \rightarrow M^{2+}(g) + e^{-}$

Third ionisation energy $M^{2+}(g) \rightarrow M^{3+}(g) + e^{-}$

Examination of successive ionisation energies for an unknown element enables us to deduce which group the element is in. We know that successive ionisation energies increase as outer electrons are removed, and that a big jump occurs when an electron is removed from a new inner orbital closer to the nucleus.

Try this yourself

(14) The graph below shows successive ionisation energies for an element **Q**. In which group of the periodic table does **Q** occur?

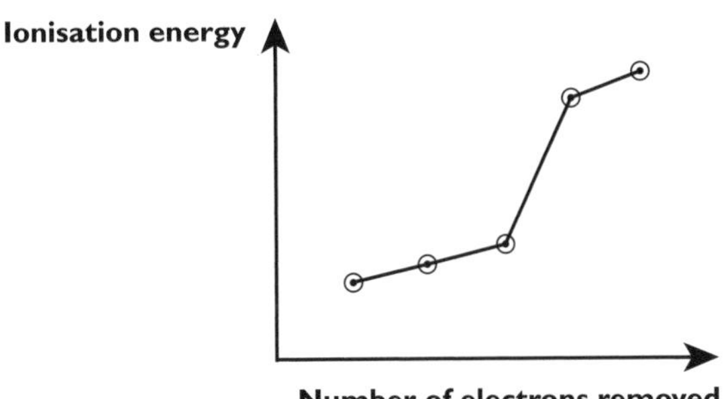

Electron affinity

Electron affinity is sometimes regarded as the reverse of ionisation energy. It is defined as the energy change for the addition of one electron to each of one mole of atoms in the gas phase.

$X(g) + e^- \rightarrow X^-(g)$

3 Chemical bonding

Chemical reactions depend on the breaking of existing bonds and the formation of new bonds. In order to understand this process, you need to be aware of the different types of bonds and forces between atoms and molecules.

Ionic (electrovalent) bonding

Ions are formed when atoms react and lose or gain electrons. Metals usually lose electrons to form positively charged **cations** — for example sodium forms Na^+. Hydrogen also loses its electron to form H^+; the ammonium ion, NH_4^+, is another example of a non-metallic cation.

Non-metallic elements gain electrons to form negatively charged **anions** — for example, chlorine forms Cl^-. Groups of atoms, such as the nitrate ion, NO_3^-, also carry negative charges.

In forming cations or anions, the elements tend to lose or gain outer electrons to attain the electron configuration of the nearest noble gas, since this is very stable. We can see this when sodium reacts with chlorine to form sodium chloride:

Na	+	Cl	→	Na^+	+	Cl^-
2,8,1		2,8,7		2,8		2,8,8

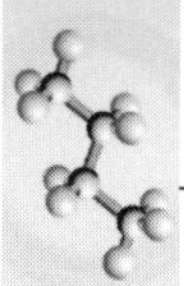

Note that 2,8 is the electron configuration of neon, and 2,8,8 that of argon, the two noble gases nearest in electron configuration to sodium and chlorine respectively.

Try this yourself

(15) Using your knowledge of the periodic table, predict the charges and electronic configuration of the ions formed by the elements in the following table.

Element	Charge on the ion	Electron configuration
Magnesium		
Lithium		
Oxygen		
Aluminium		
Fluorine		
Sulfur		

How do we know that ions exist?

The evidence for the existence of ions comes from electrolysis. An electric current can be passed through a molten salt or an aqueous solution of the salt (Figure 3.1). This relies on the movement of ions in the solution carrying the charge, followed by the loss or gain of electrons at the appropriate electrode to form elements.

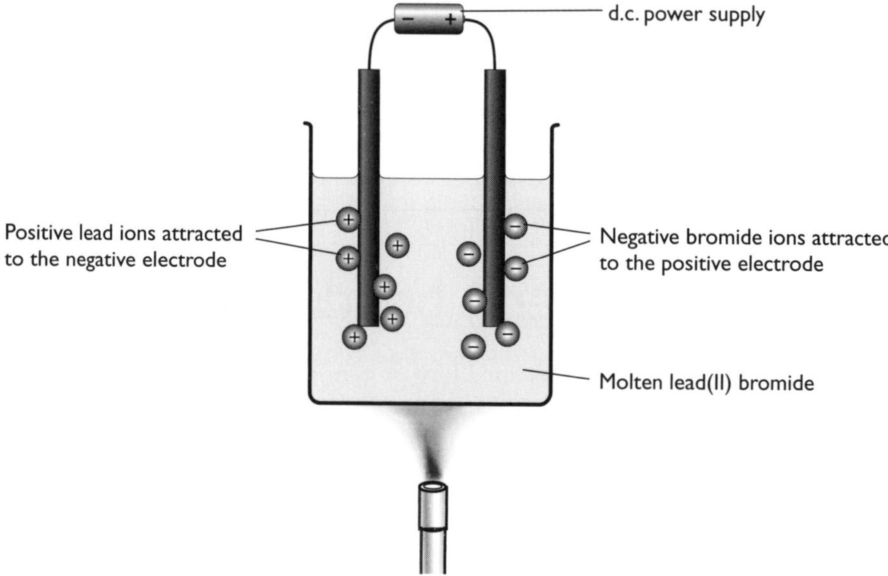

Figure 3.1 Electrolysis

Ionic crystals

In the solid state (see also Chapter 4), cations and anions come together to form ionic crystals (Figure 3.2). These consist of a giant three-dimensional lattice of ions. The structure of these crystals depends on the relative sizes of the anion and cation, and on the stoichiometry of the compound concerned.

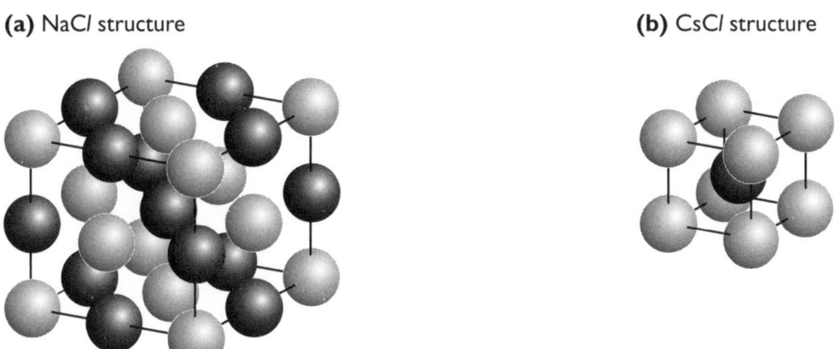

Figure 3.2(a) Crystal structure of sodium chloride (b) Crystal structure of caesium chloride

The three-dimensional structure within the crystal is held together by the net attractive forces between the oppositely charged ions. There are also longer-range repulsive forces between ions of the same charge, but because these are longer range they are weaker.

Covalent and coordinate (dative) bonding

The major difference between ionic (electrovalent) and covalent bonding is that in ionic (electrovalent) bonding electrons are *transferred* from one element to another to create charged ions, while in covalent bonding the electrons are *shared* between atoms in pairs.

It is important to remember that the electrons do not 'circle around the nucleus', but exist in a volume of space surrounding the nucleus where there is a high probability of finding the electron. These are known as orbitals. A covalent bond is formed due to the overlap of orbitals containing electrons and the attraction of these bonding electrons to the nuclei of both atoms involved.

It is not necessary to have atoms of different elements to form covalent bonds, so it is possible for an element to form molecules that have a covalent bond between the atoms, e.g. chlorine, Cl_2 (Figure 3.3).

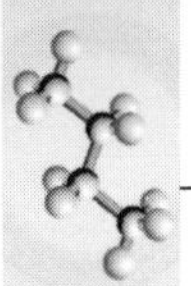

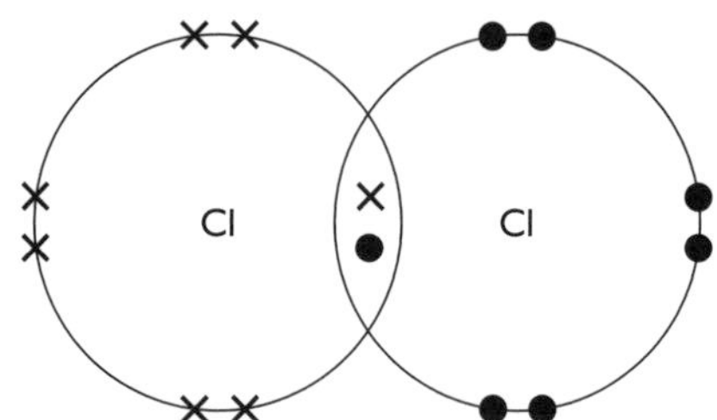

Figure 3.3 Covalent bonding in chlorine

Dot-and-cross diagrams

These are diagrams that represent the movement of electrons in the formation of both electrovalent and covalent bonds, with the electrons from one atom represented by 'dots' and those from the other atom by 'crosses'. The electrons are, of course, identical, but this system helps to visualise what is happening when the bond is formed.

In an electrovalent bond, one or more electrons are transferred from one element (usually a metal) to another element (usually a non-metal). The transfer in the formation of magnesium oxide is shown in Figure 3.4.

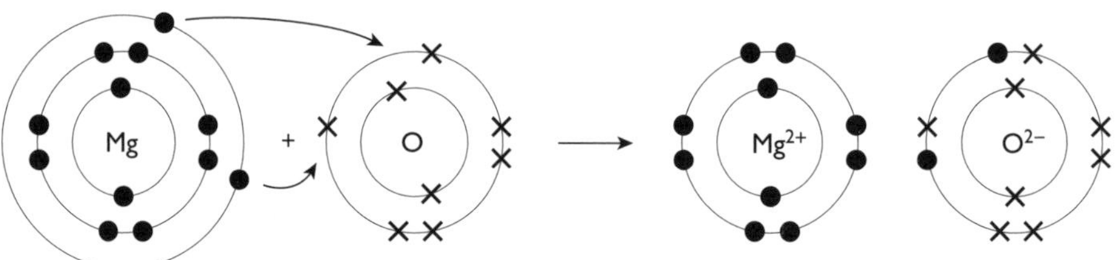

Figure 3.4 Electron transfer in the formation of magnesium oxide

Try this yourself

(16) Use a copy of the periodic table to help you draw dot-and-cross diagrams for the following:

- **(a)** hydrogen, H_2
- **(b)** water, H_2O
- **(c)** carbon dioxide, CO_2
- **(d)** methane, CH_4
- **(e)** lithium fluoride, LiF

The bonded atoms in a covalent bond usually have a 'share' of an octet of electrons associated with each atom, but this is not always the case. For example in boron trichloride, BCl_3, there are only six electrons associated with the boron atom (Figure 3.5).

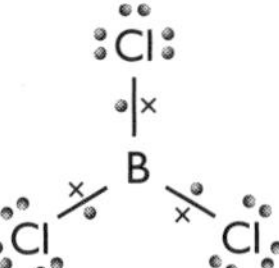

Figure 3.5 Boron trichloride

Coordinate or dative covalent bonds are formed when both electrons in a pair come from the same atom, for example in NH_4^+ (Figure 3.6). Once formed, the bond is not distinguishable from the other covalent bonds in the ion.

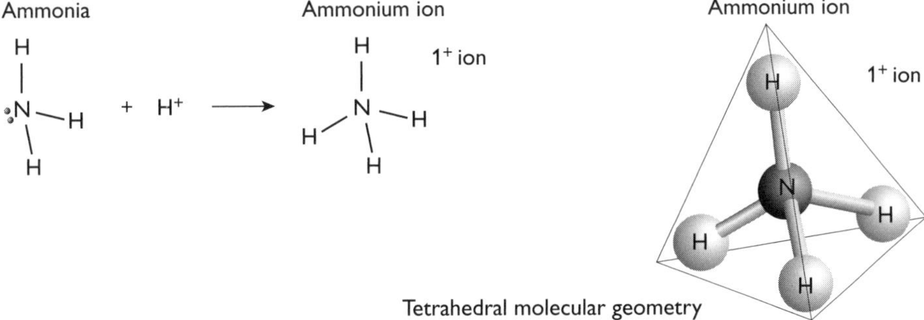

Figure 3.6 The ammonium ion

There are plenty of other examples of this type of covalent bonding — for example, in carbon monoxide, CO, and in the nitrate ion, NO_3^-, and particularly in the formation of transition metal complexes (see Chapter 11).

It is possible to have multiple covalent bonds, depending on the number of pairs of electrons involved. This can occur in simple molecules such as oxygen, O_2, (Figure 3.7(a)) but is particularly important in carbon compounds such as ethene, C_2H_4 (Figure 3.7(b)). (See also Chapter 16).

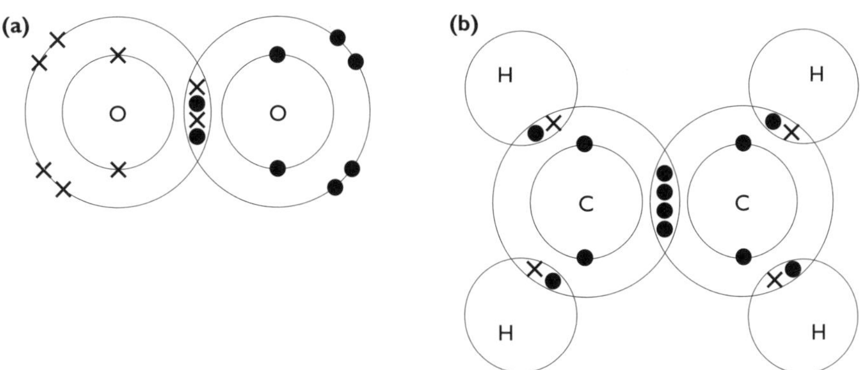

Figure 3.7(a) Multiple covalent bonding in (a) oxygen and (b) ethene

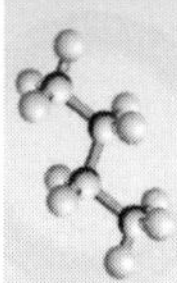

Simple molecular shapes

Unlike ionic (electrovalent) bonds that have no particular direction, covalent bonds are directional. This means that covalently bonded molecules have distinctive shapes depending on the number of bonds, since the pairs of electrons in the bonds repel any other pairs. Figure 3.8 shows the basic shapes that molecules containing up to four electron pairs can adopt.

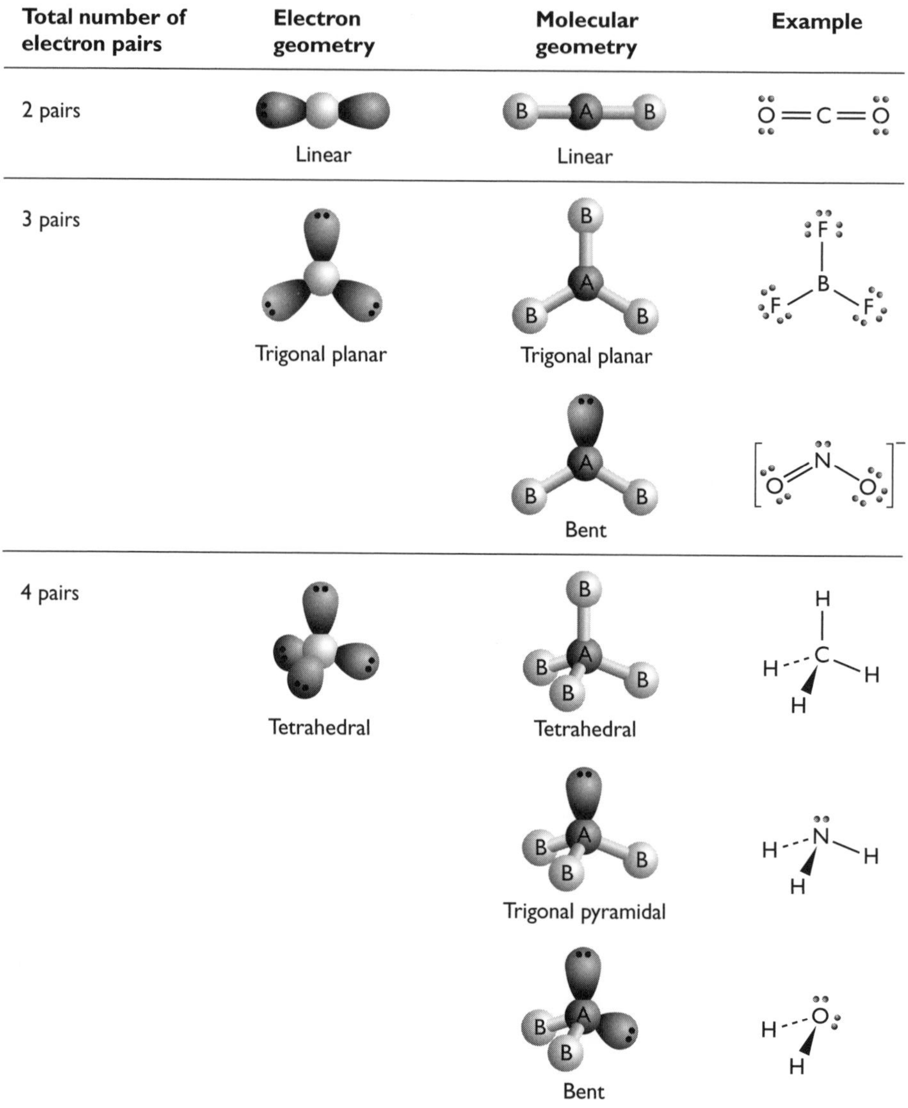

Total number of electron pairs	Electron geometry	Molecular geometry	Example
2 pairs	Linear	Linear	O=C=O
3 pairs	Trigonal planar	Trigonal planar	
		Bent	
4 pairs	Tetrahedral	Tetrahedral	
		Trigonal pyramidal	
		Bent	

Figure 3.8 Shapes of molecules

In some circumstances more than four pairs of electrons can be involved, as in the case of sulfur hexafluoride, SF_6. The repulsion effect still applies. In this case the molecule is octahedral.

In addition, any non-bonded pairs of electrons will repel bonded pairs, but occupy rather more space. This means that the bond angles are widened. We can see this effect if we compare three molecules, each with four pairs of electrons — methane, CH_4, ammonia, NH_3, and water, H_2O. The normal tetrahedral angle is 109.5°, but ammonia has a lone pair of electrons that squeezes the H—N—H bond angle to 107°. In water, the two lone pairs of electrons present squeeze the H—O—H bond angle even more, reducing it to 104.5°. This is shown in Figure 3.9.

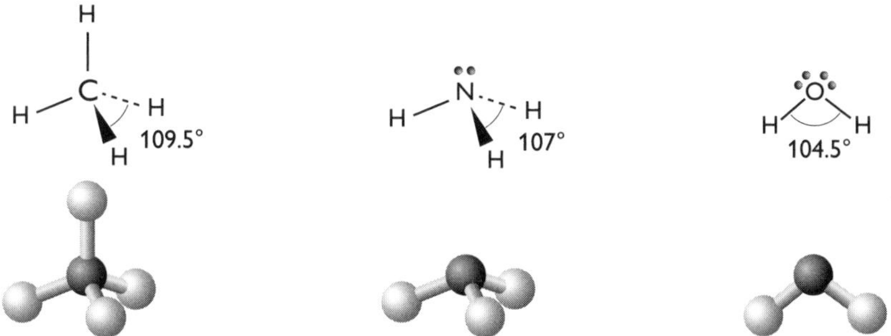

Figure 3.9 Bond angles in methane, ammonia and water

Giant molecular structures

As well as forming simple molecules like those shown on the page opposite, it is possible to form giant molecular structures. In this syllabus, these are confined to different structural forms of carbon (diamond and graphite) and silicon dioxide, which is similar to diamond. Examples of these are shown on page 46.

Bond energies, bond lengths and bond polarities

When two atoms join to form a covalent bond the reaction is exothermic — energy is given out. It therefore follows that to break that covalent bond energy must be supplied. The **bond energy** is defined as the average standard enthalpy change for the breaking of one mole of bonds in a gaseous molecule to form gaseous atoms:

Br—Br(g) + bond energy → 2Br(g)

It follows from this that bond energy is an indication of the strength of the forces holding the atoms together in a covalent molecule.

Bond energies can vary from around 150 kJ mol^{-1} for molecules with weak bonds, such as I—I, to 350–550 kJ mol^{-1} for stronger bonds, such as C—C and O—H, to around 1000 kJ mol^{-1} for very strong bonds such as N≡N. Bond energy increases with the number of electron pairs making up the bond. Thus, E(C—C) = 350 kJ mol^{-1}; E(C=C) = 610 kJ mol^{-1} and E(C≡C) = 840 kJ mol^{-1}.

Bond length is defined as the distance between the middle of the atoms at either end of the bond. The length of a bond depends on a number of factors, particularly the number of pairs of electrons making up the bond. So for the three carbon bonds

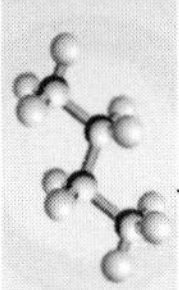

described above, the bond lengths are C—C, 154 pm; C═C, 134 pm and C≡C, 120 pm (1 pm is 1 picometre or 1×10^{-12} metres).

Since most covalent bonds are between different atoms, and different nuclei have a different attraction for the electrons, it follows that the electrons in a covalent bond are pulled closer to one atom than the other. The degree of attraction depends on the nature of the two atoms involved. This effect of unequal sharing of electrons is called **bond polarity**. The measure of this tendency to attract a bonding pair of electrons is called **electronegativity**. You need to remember that:

- electronegativity increases from left to right across a period
- electronegativity decreases down a group
- small atoms with many protons in the nucleus have high electronegativity
- the greater the difference in electronegativity of the two atoms, the more polar the bond will be

Orbital overlap: σ-bonds and π-bonds

At the start of the section on covalent bonds we talked about them being formed by the overlap of electron orbitals on each atom. Most covalent bonds are found in compounds of carbon, and it is important to understand how such bonds are formed.

In carbon, the 2s- and 2p-orbitals are quite close in energy. This means that it is possible to promote one of the 2s-electrons to the empty 2p-orbital. The energy required for this promotion is more than compensated for by the energy released when four bonds are formed (compared with the two bonds that could have been formed from the two 2p-orbitals that each contained a single electron). This can be seen in Figure 3.10.

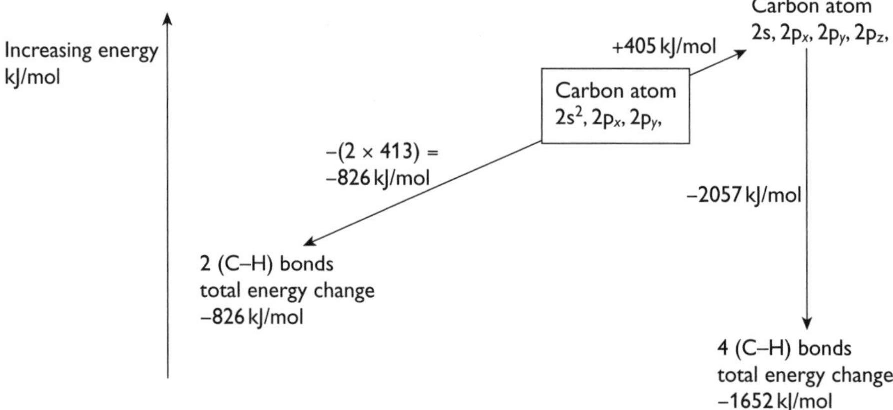

Figure 3.10 Energy benefit in forming four carbon–hydrogen bonds

The four electrons form four identical orbitals that have some s and some p characteristics. These are known as sp^3 hybrid orbitals. In forming methane, CH_4, they overlap with the s-orbitals of hydrogen atoms. Bonds formed from the overlap of orbitals with some s character are called **sigma bonds** (σ-bonds). Bonds formed by the overlap of p-orbitals are called **pi-bonds** (π-bonds).

There are two other ways in which the orbitals on carbon can be hybridised. First the s-orbital may be hybridised with two of the p-orbitals to form three sp^2-orbitals leaving the remaining 2p-orbital unchanged. The three sp^2-orbitals lie in a plane 120° apart, with the 2p-orbital at right angles to this. This is the type of hybrid orbital formed by the carbon atoms in ethene and benzene.

Look at the structure of ethene. One pair of sp^2-orbitals overlap forming a σ-bond. This brings the 2p-orbitals on the two carbons close enough together for them to overlap forming a π-bond. The bonding in ethene can be seen in Figure 3.11.

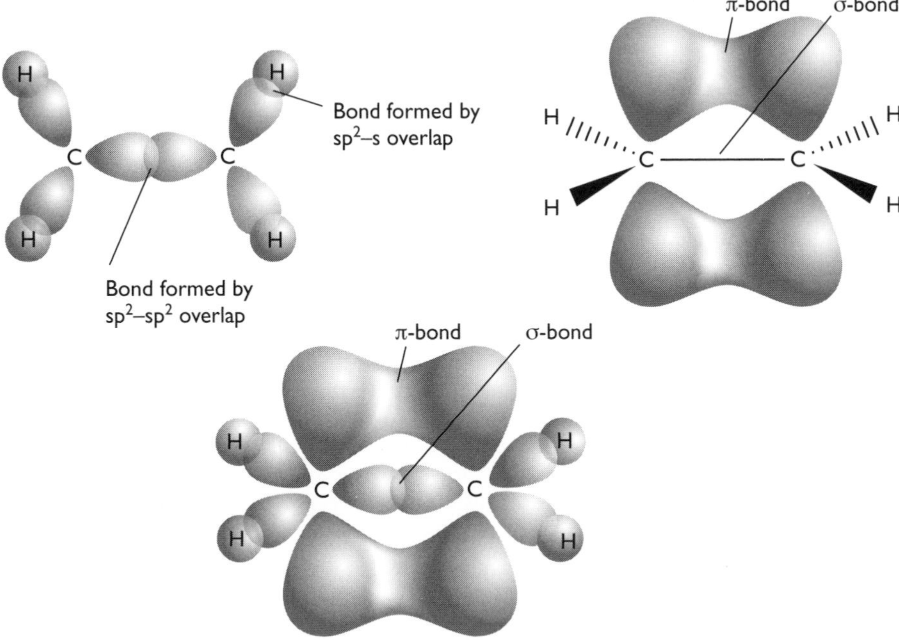

Figure 3.11 Bonding in ethene

A similar bonding pattern occurs in molecules of benzene, C_6H_6. However, in benzene the carbon atoms are arranged in a hexagonal ring. The 2p-orbitals overlap above and below the ring forming circular molecular orbitals, as shown in Figure 3.12. The electrons are said to be **delocalised** since they no longer belong to individual carbon atoms.

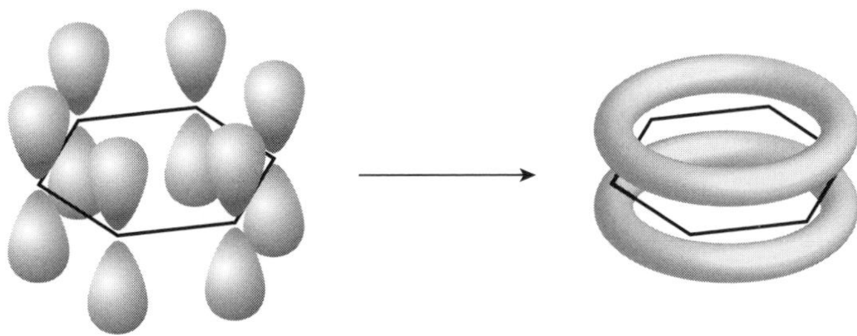

Figure 3.12 Bonding in benzene

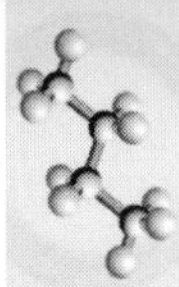

Intermolecular forces

As well as ionic (electrovalent) and covalent bonding there are a number of other forces that exist between molecules. These are:

- van der Waals forces
- permanent dipole–dipole interactions
- hydrogen bonds

van der Waals forces

These are the weakest of the forces. They act between all particles whether they are polar or non-polar. They exist due to the movement of electrons that in turn causes instantaneous dipoles. These induce dipoles in neighbouring molecules.

Permanent dipole-dipole interactions

These occur between polar covalent molecules, i.e. those containing different elements. An example is the forces between the dipoles in a molecule such as H–C*l* (Figure 3.13).

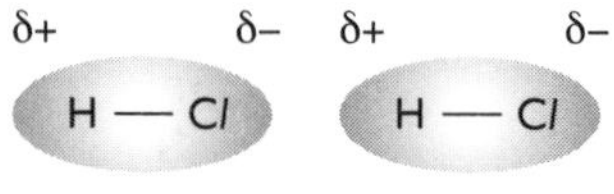

$\delta+$ H — C*l* $\delta-$ $\delta+$ H — C*l* $\delta-$

Figure 3.13 Dipole–dipole interactions in hydrogen chloride

Hydrogen bonds

This is a particular sort of comparatively strong dipole–dipole interaction between molecules containing hydrogen with nitrogen, oxygen or fluorine. These bonds result from the lone pairs of electrons on the nitrogen, oxygen or fluorine atoms, so the hydrogen atom can be considered as acting as a 'bridge' between two electronegative atoms.

This form of bonding can have significant effects on the physical properties of the compound concerned. For example, based on its molecular mass, water would be expected to exist as a gas at room temperature. The fact that it exists as a liquid at room temperature is due to the hydrogen bonding present (Figure 3.14). As a result of hydrogen bonding, water possesses surface tension, which enables some insects to walk on its surface. Finally, the fact that ice is less dense than liquid water and floats on the surface is also a result of hydrogen bonding.

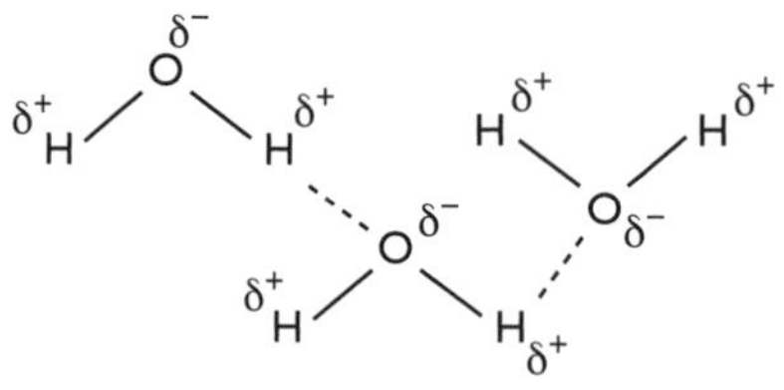

Figure 3.14 Hydrogen bonding in water

Metallic bonding

Metals have distinctive properties, many of which are based on the fact that metals possess a regular lattice of atoms, in much the same way as an ionic (electrovalent) crystal. The main difference is that all of the atoms in a metallic lattice are the same and the outer electrons are not held by the atoms but are delocalised throughout the lattice (Figure 3.15). It is these mobile electrons that give metals their electrical conductivity.

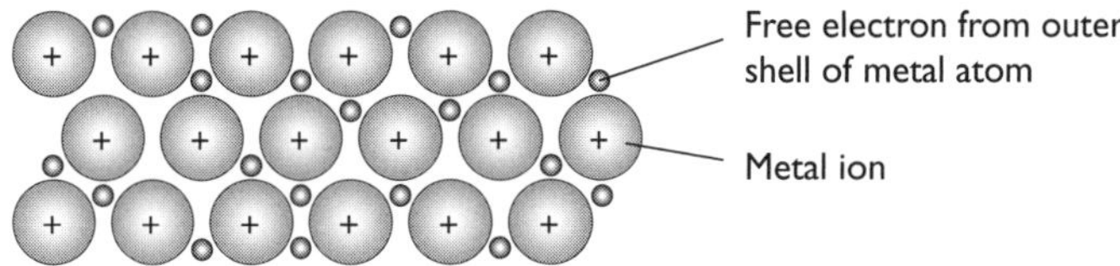

Figure 3.15 Metallic bonding

Try this yourself

(17) For each of the materials in the table, predict the main type(s) of intermolecular forces that exist in the material.

Material	**Intermolecular force**
Methanol, CH_3OH	
Magnesium oxide, MgO	
Iodine chloride, IC*l*	
Argon, Ar	
Aluminium, A*l*	

Bonding and physical properties

The type of bonding in a substance affects its physical properties. Ionic (electrovalent) compounds, which are formed of giant lattices of oppositely charged ions, tend to have high melting and boiling points; they usually dissolve in water and they conduct electricity when molten. Covalently bonded compounds tend to be gases, liquids or low melting point solids; they dissolve in covalent solvents and are electrical insulators. Metals have a giant lattice structure with a 'sea' of mobile electrons. In general, metals have high melting points, can be bent and shaped, and are good electrical conductors.

4 States of matter

All substances exist in one of the three states of matter — gas, liquid or solid. At AS, you need to know the theories concerning particles in a gas, together with the forces

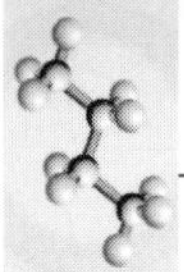

between particles in liquids and solids and how these influence the properties of the substances concerned.

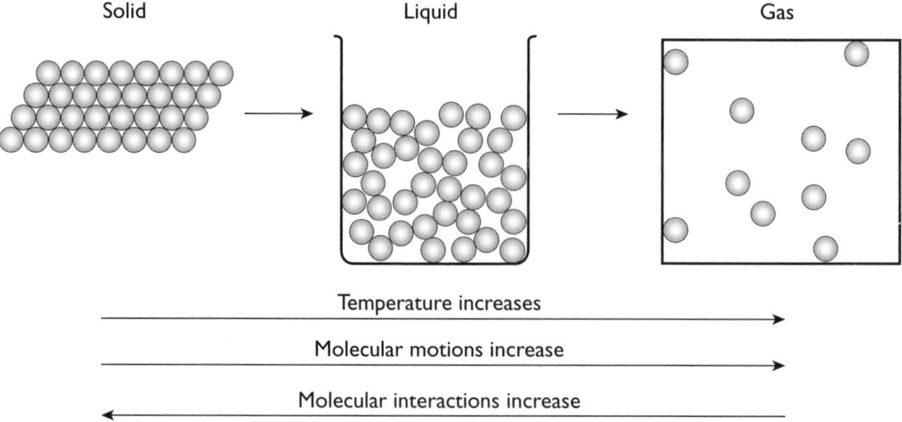

Figure 4.1 States of matter

The gaseous state

It helps to have some simple ideas about gases:

- Gases have a very low density because the particles are spaced widely apart in the container.
- Gases are easily compressed because of the large spaces between the particles.
- Gases have no fixed shape or volume and, because of the lack of particle attraction, they spread out and fill the container.

The rapid and random movement of the particles in all directions means that gases readily 'spread out' or **diffuse**. The overall movement of the particles is in the direction of lower concentration from a higher concentration. This is why you can smell perfume or food (or less pleasant smells!) when the source is some distance from us.

Bearing these properties in mind, let's look in more detail at how gas particles interact both with each other and with the walls of a container.

Ideal gases

We can try to explain the physical properties of gases by imagining the particles in constant random motion colliding both with each other and with the walls of the container. This idea of an **ideal gas** is based on certain assumptions:

- The volume of the particles themselves is negligible compared with the volume of the container.
- The particles are not attracted to each other, or to the walls of the container.
- All collisions are perfectly elastic, so there is no change in the kinetic energy of the particles.
- The particles are in continuous motion colliding frequently with each other and with the walls of the container.

These assumptions lead to the following properties of the gas:

- the bombardment of the container walls by the particles causes pressure
- the average kinetic energy of the particles is directly proportional to the absolute temperature on the Kelvin scale (K).

Two laws related to gases under ideal conditions were developed by two scientists, Boyle and Charles.

Boyle's law states that for a given mass of gas at a constant temperature, the product of the pressure(p) and the volume(V) is a constant:

$$pV = \text{constant}$$

Charles' law can be stated a number of ways, but one is: at constant pressure, for a given quantity of gas, the ratio of its volume (V) and the absolute temperature (T) is a constant:

$$\frac{V}{T} = \text{constant}$$

These two laws can be combined to give the **ideal gas equation**:

$$pV = nRT$$

where n is the number of moles and R is a constant known as the universal molar gas constant.

It is vital to make sure that you use consistent units when using this equation. The most common units are shown in Table 4.1.

Table 4.1

Pressure (p)	Volume (V)	Number of moles (n)	Gas constant (R)	Temperature (T)
Pa (pascals)	m^3	mol	$8.314\,J\,mol^{-1}\,K^{-1}$	K
atm	dm^3	mol	$0.08206\,atm\,dm^3\,mol^{-1}\,K^{-1}$	K

In examinations, you might be asked to use the ideal gas equation to determine the M_r of a compound. Rearranging the equation gives:

$$n = \frac{pV}{RT}$$

In a known mass of gas, m, the number of moles present is m/M_r. Substituting this for n gives:

$$M_r = \frac{mRT}{pV}$$

Example

In an experiment, 0.217 g of a liquid was vaporised in a syringe placed in an oven at 80 °C and 101 000 Pa pressure. The vaporised liquid gave 66.0 cm^3 of gas. Calculate the relative molecular mass of the liquid.

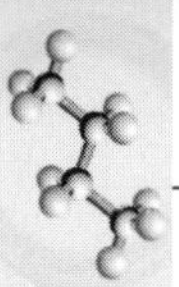

Answer

Use $M_r = \dfrac{mRT}{pV}$

since p is in Pa, R must be 8.314.

Substituting gives $M_r = \dfrac{0.217 \times 8.314 \times 353}{101\,000 \times 6.6 \times 10^{-5}} = 95.2$

Note that the volume has been converted from cm^3 to dm^3 and the temperature of 80 °C to 353 K.

Non-ideal (real) gases

With real gases, some of the assumptions made about ideal gases no longer apply under all conditions. At very high pressure particles are closer together. At low temperature, the particles move less rapidly. Under both of these conditions, the volume of the particles becomes significant. Since we know that gases can be liquefied under these conditions, it follows that there must be forces of attraction between the particles. In an ideal gas, pV is a constant and a plot of pV against p would be expected to be a horizontal line. Figure 4.2 shows the result for a number of real gases.

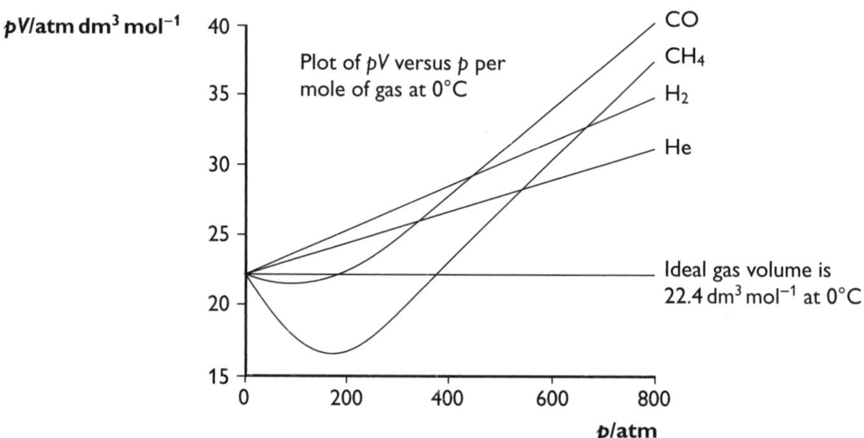

Figure 4.2 Non-ideal gas behaviour

It follows that for a real gas to behave close to ideality, it must be at low pressure and/or high temperature. In addition, it should have a low M_r and weak forces between the particles. (Note that in Figure 4.2, helium is the line closest to ideal.)

The liquid state

Liquids are very different from gases and properties of liquids are closer to those of solids than to those of gases. This is not too surprising if you consider the difference in the gaps between particles in the three different states. This difference means that the intermolecular forces are much more like those in solids than the very weak forces in gases. Figure 4.1 shows the difference in particle arrangements in the three states.

Some familiar properties of liquids are:

- Liquids have a much greater density than gases because the particles are much closer together, resulting in attractive forces between them.
- Liquids are not readily compressed because of the lack of space between the particles.
- Liquids have a surface and a fixed volume (at constant temperature) due to the increased particle attraction; the shape is not fixed and is determined by the container.

The random movement of the particles means that liquids can diffuse, but the diffusion is much slower in liquids compared with in gases. This is because there is less space for the particles to move in and more collisions occur, slowing down the diffusion process.

Melting

This is the term given to the change of a solid into a liquid. If you think about the bonds between the particles in solids, these bonds need to be broken in order for the solid to melt — in other words energy must be supplied. The particles separate but do not move very far apart, so the energy required depends only on the strength of the original bonds in the solid lattice. This is illustrated in Figure 4.3.

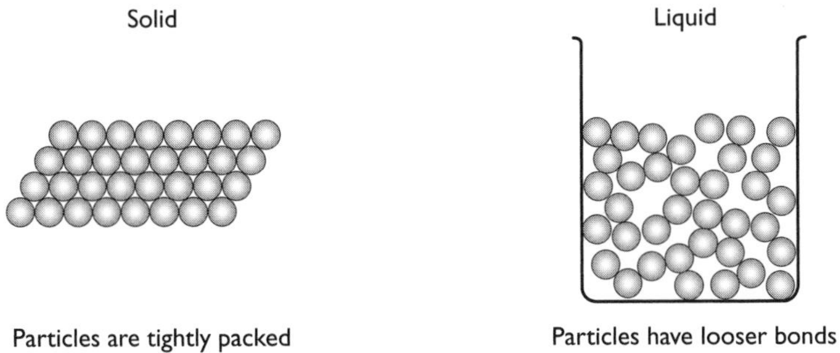

Figure 4.3 Melting point

Vaporisation

The changing of a liquid to a gas is called vaporisation (see Figure 4.4). This requires energy because the bonds between the particles, which are close together in the liquid, need to be broken. The particles must also be given enough energy to be separated by considerable distances.

Boiling is rapid evaporation anywhere in the bulk liquid at a fixed temperature called the boiling point. It requires continuous addition of heat. The rate of boiling is limited

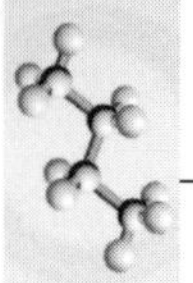

by the rate of heat transfer into the liquid. Evaporation takes place more slowly than boiling at any temperature between the melting point and boiling point, and only from the surface, and results in the liquid becoming cooler due to loss of particles with higher kinetic energy.

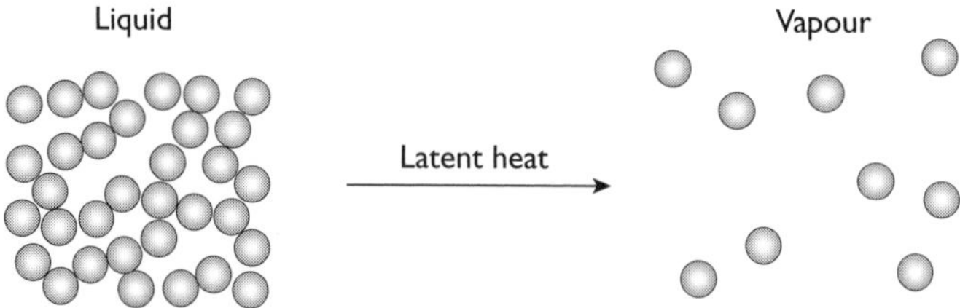

Figure 4.4 Vaporisation

Heating and cooling 'curves'

When a solid is heated so that first it melts and the liquid then vaporises, there are two changes of state. A graph of temperature against time (or energy supplied) for this process has an interesting shape (Figure 4.5).

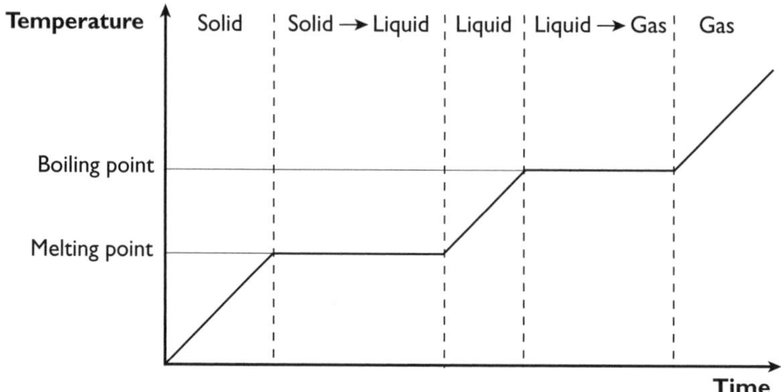

Figure 4.5 Heating curve

The two horizontal portions show that when the state of a substance changes the energy absorbed goes into weakening the bonds between particles and so there is no temperature rise until this process is complete. The reverse is true when a hot gas is cooled, with energy released as bonds form between the particles. This is the reason that steam produces a much worse 'burn' than a small amount of boiling water at the same temperature.

The solid state

We have already looked at some ideas concerned with bonding in Chapter 3, so you are familiar with the types of bonding in solids:

- ionic (electrovalent)
- covalent — simple and giant
- hydrogen
- metallic

To answer questions from this section of the syllabus you need to be able to describe simply a selection of crystalline solids. To do this, it is usually easiest to make a sketch and then add notes. These diagrams do not need to be as complex as those in a textbook, but they must show the important points (see Figure 4.6).

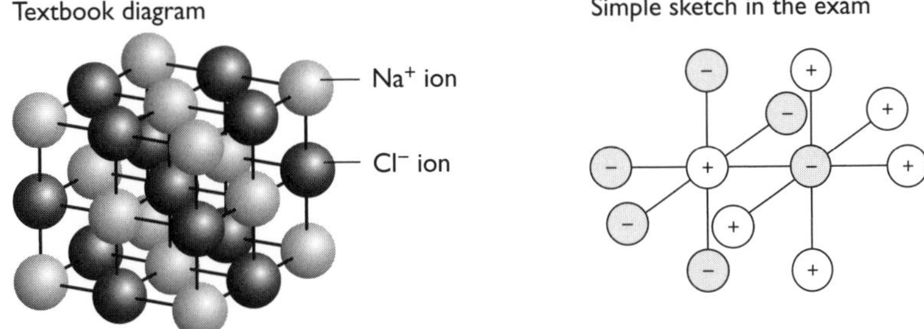

Figure 4.6 Comparison between a textbook diagram and a simple sketch

The notes should focus on the physical properties of the solid in terms of melting point, solubility and conductivity. Table 4.2 should help you here.

Table 4.2

	Melting point	**Solubility**	**Electrical conductivity**
Electrovalent, e.g. sodium chloride, magnesium oxide	High melting point due to strong electrostatic forces between ions	Soluble in polar solvents because ions interact with water molecules and the lattice breaks down	Insulator when solid, conductor when molten or when in solution because the ions are mobile
Simple covalent, e.g. iodine	Low melting point due to weak forces between the molecules	Insoluble in polar solvents but may dissolve in non-polar solvents	Insulator — no ions or free electrons to conduct
Giant covalent, e.g. diamond, graphite, silicon(IV) oxide	Very high melting point due to strong covalent bonds forming the giant lattice	Mostly insoluble in both polar and in non-polar solvents	Mostly insulators due to electrons localised in bonds (graphite is an exception due to delocalised electrons in its lattice)
Hydrogen-bonded, e.g. ice	Low melting point due to relatively weak forces between the molecules	Soluble in polar solvents because hydrogen bonds form with the solvent	Insulator — no ions or free electrons to conduct
Metallic, e.g. copper	Relatively high melting point because of strong metallic bonds	Insoluble in water but may react with water	Conducts when solid due to delocalised electrons in lattice

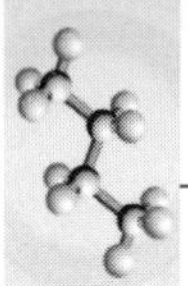

Tip Exam questions may ask you to use your knowledge of bonding and structure to explain the properties of a given material. In answering such questions it is important to know both the type of structure you are dealing with and the types of bond present.

Ceramics

Ceramics are an interesting group of materials since they are in such common use both industrially and domestically. Ceramics are hard, brittle, heat-resistant and corrosion-resistant materials made by shaping and then firing a non-metallic mineral, such as clay, at a high temperature. The ceramics that are of interest at AS are those based on magnesium oxide, aluminium oxide and silicon(IV) oxide. These are used to make furnace linings, electrical insulators such as those found on high-voltage power lines and materials such as glass and crockery.

Example

Explain the strength, high melting point and electrical insulating properties of ceramics in terms of their giant molecular structure.

Answer

There are enough clues in the question to give you a good start at answering it. You have been told what the structure is (giant molecular) and which properties to focus on (strength, high melting point and electrical insulating properties). So if you look at Table 4.2, you can see that the ceramics:

- have very high melting points because of strong covalent bonds forming the giant lattice
- are insulators because of electrons localised in bonds
- are strong because of the strong covalent bonds in the lattice

The two diagrams in Figure 4.7 show the structures of diamond and silicon(IV) oxide. By looking at the two structures and knowing the properties of diamond, you should be able to predict properties of silicon(IV) oxide.

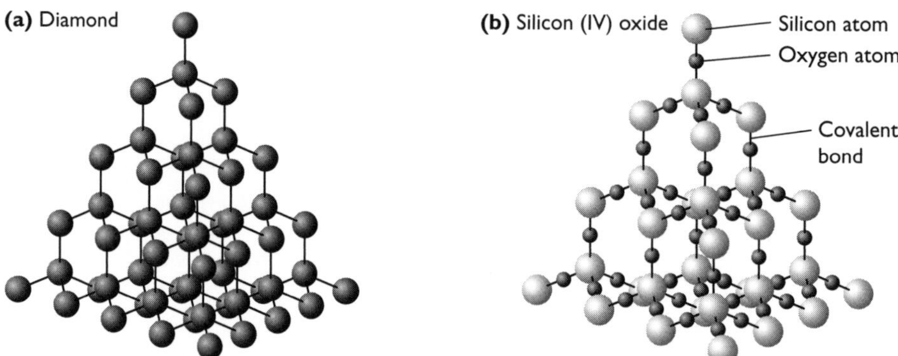

Figure 4.7 Structure of (a) diamond and (b) silicon(IV) oxide

Metal alloys

Metals are of interest not just in their pure state, but also because some can form alloys — some atoms of one metal are substituted for some atoms in the lattice of another metal (substitutional alloys). The alloy has different properties from the original metals, for example:

- colour (as in the case of zinc and copper forming brass)
- hardness (again with brass)

Another example is the use of a non-metal (carbon) as well as other metals to produce the variety of specialist steels now available. Aluminium alloys have become increasingly important as the demand for lightweight, strong materials has increased. Figure 4.8 shows the difference between a typical alloy, such as brass, and a steel (interstitial alloy), where the small carbon atoms fit between metal atoms.

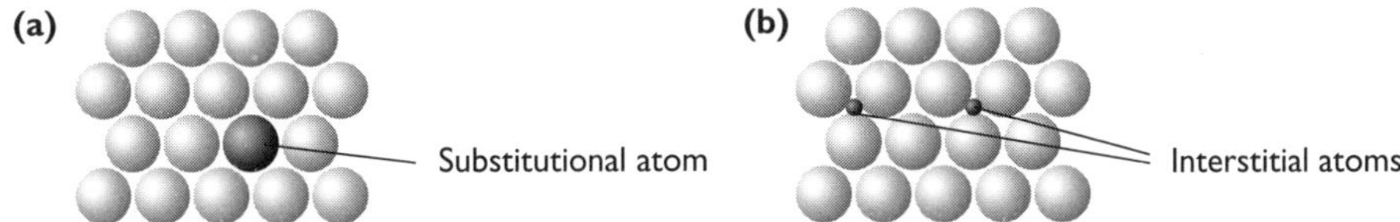

Figure 4.8 Structure of (a) a typical alloy and (b) a steel

Metals have to be mined and separated from the waste rock. Today, most of the ores rich in metals have already been mined and we are using ores that contain much less metal. Under these circumstances it makes sense to try to recycle metals because new extraction from low-grade ores requires huge amounts of energy.

Try this yourself

(18) Study the data given about material **X.** Suggest, with reasons, the type of structure and bonding present in **X**.

- **X**, a soft waxy solid, melts at just under 100 °C.
- It is an electrical insulator, both as a solid and when molten.
- It dissolves in cyclohexane to give a solution that does not conduct electricity.

Ice and water

Water is an extremely unusual compound. For example, from its relative molecular mass (18) we would expect it to be a gas (compare with nitrogen (28), oxygen (32) and carbon dioxide (44)). It is the presence of relatively strong hydrogen bonds that make it a liquid. Without these, life would not exist. The boiling points of some hydrogen compounds are compared in Figures 4.9(a) and (b).

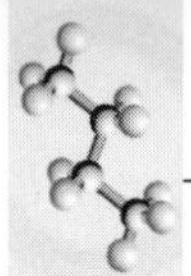

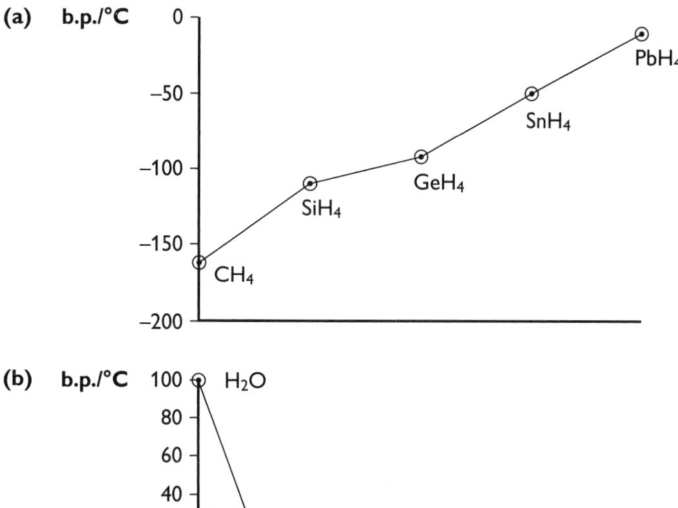

Figure 4.9 A comparison of the boiling points of some hydrogen compounds

Many elements form compounds with hydrogen called **hydrides**. The boiling points for the hydrides of Group IV elements are plotted in the graph shown in Figure 4.9(a). Graphs for the hydrides of elements in Groups V, VI and VII are very different (Figure 4.9(b)). Although for the second and subsequent elements in each group you get the expected pattern, the first element in each group shows a much higher boiling point than expected. The reason for this is that these compounds have significantly stronger bonds between the particles. These are **hydrogen bonds**, which were discussed in Chapter 3.

Another unusual property of water is that its solid form (ice) is less dense than the liquid form. When water freezes, the random orientation of the particles in the liquid changes gradually as hydrogen bonds form. This produces an open structure with gaps in the lattice making the solid less dense than the liquid state (Figure 4.10).

The presence of hydrogen bonds in liquid water also helps it to be a good solvent for ionic (electrovalent) substances. The partially-negative oxygen atom is attracted to cations and the partially-positive hydrogen atoms are attracted to anions (Figure 4.11).

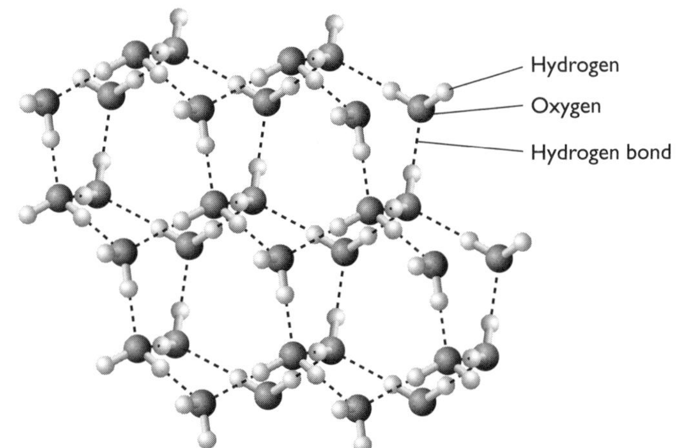

Figure 4.10 Structure of ice

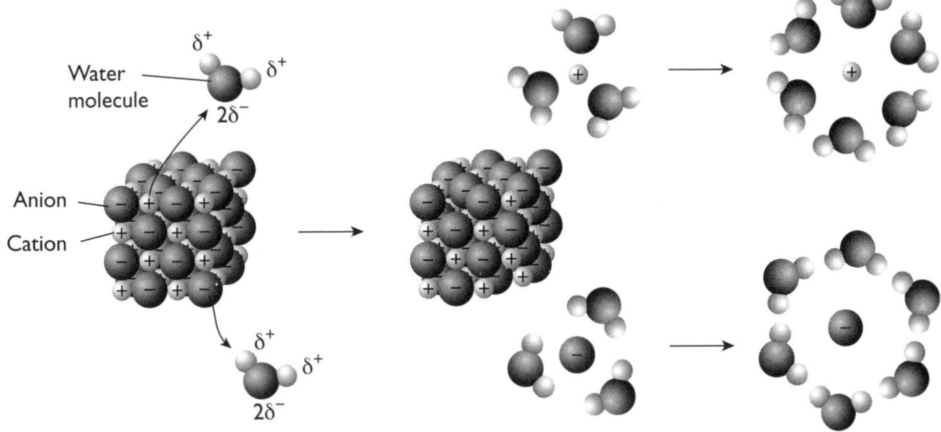

Figure 4.11 Water as a solvent

5 Chemical energetics

Almost all chemical reactions are accompanied by energy changes as bonds are broken and/or formed. Usually the energy changes involve heat, but they can also involve sound, light or even electrical energy.

Try this yourself

(19) Can you think of examples of chemical reactions that give out or use:

(a) sound

(b) light

(c) electrical energy?

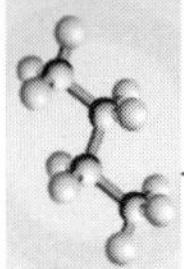

We are most familiar with reactions that give out heat — a test tube gets warmer or a fuel is burned. These are called **exothermic** reactions. A smaller number of reactions take in energy overall, and these are known as **endothermic** reactions. The overall energy changes in these two types of reaction are shown in Figure 5.1.

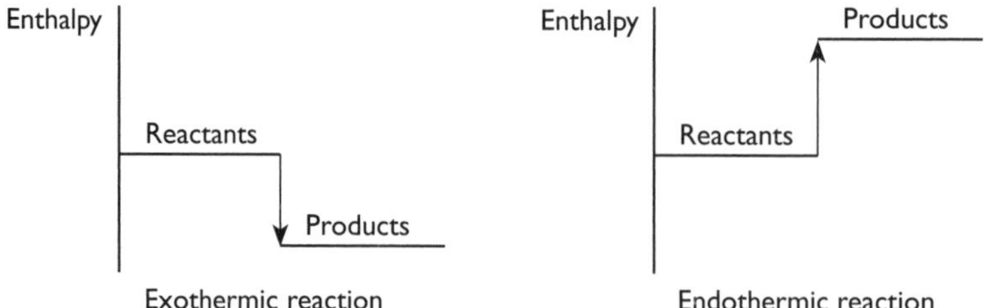

Figure 5.1 Energy change in an exothermic and an endothermic reaction

The vertical (y) axes in Figure 5.1 represent the enthalpy (see below) of the compounds.

So you can see that in an exothermic reaction the enthalpy change is in a negative direction. This is worth remembering since exothermic reactions always show a negative enthalpy change. It follows that endothermic reactions show a positive enthalpy change, and this is borne out in the Figure 5.1. Enthalpy changes are measured in $kJ\,mol^{-1}$.

In order to make sense of what takes place in a chemical reaction, we need to consider **standard conditions**. Using standard conditions means that the results of measurements are reproducible.

Standard conditions are:
- all reagents and products are in their most stable state
- the pressure is 1 atmosphere
- the temperature is specified (usually 298 K, 25°C)

Enthalpy changes

In textbooks you will see references made both to *energy changes* and to *enthalpy changes*, and it is important to understand the difference in the way these are used in questions and in the syllabus. Enthalpy changes always refer to particular sets of conditions, at A-level these are usually referred to as *standard enthalpy changes*, which means they refer to the heat absorbed or given out at 298 K and standard pressure (usually 1 atmosphere or 100 kPa, although some textbooks refer to 1 atmosphere as 101 kPa). The examples that follow outline the particular enthalpy changes you need to know about.

The sorts of reactions for which you may need to measure or calculate the enthalpy change are as follows:
- **formation** — the enthalpy change when 1 mole of substance is formed from its elements under standard conditions, for example:

$2Li(s) + ½O_2(g) \rightarrow Li_2O(s)$ $\Delta H^\ominus = -596\,kJ\,mol^{-1}$

- **combustion** — the enthalpy change when 1 mole of substance is completely burned in excess oxygen under standard conditions, for example:

  $CH_4(g) + 2O_2(g) \rightarrow CO_2(g) + 2H_2O(g)$ $\Delta H^\ominus = -882\,kJ\,mol^{-1}$

- **hydration** — the enthalpy change when 1 mole of gaseous ions become hydrated under standard conditions, for example:

  $Al^{3+}(g) + aq \rightarrow Al^{3+}(aq)$ $\Delta H^\ominus = -4613\,kJ\,mol^{-1}$

- **solution** — the enthalpy change when 1 mole of substance is completely dissolved, under standard conditions, in enough solvent so that no further heat change takes place on adding more solvent, for example:

  $HCl(g) + (aq) \rightarrow HCl(aq)$ $\Delta H^\ominus = -74.4\,kJ\,mol^{-1}$

- **neutralisation** — the enthalpy change when an acid is neutralised by an alkali to produce 1 mole of water under standard conditions. This can also be thought of as the enthalpy change when 1 mole of hydrogen ions reacts with 1 mole of hydroxide ions:

  $H^+(aq) + OH^-(aq) \rightarrow H_2O(l)$ $\Delta H^\ominus = -57.1\,kJ\,mol^{-1}$

- **atomisation** — the enthalpy change for the formation of 1 mole of gaseous atoms from an element under standard conditions, for example:

  $½O_2(g) \rightarrow O(g)$ $\Delta H^\ominus = 249\,kJ\,mol^{-1}$

Another enthalpy change you need to understand is **bond energy**, which was described in Chapter 3. This was talked about as the energy needed to *break* a bond, so it was energy supplied. This means that ΔH is positive.

The final enthalpy change you need to be aware of for A2 is **lattice energy**. You need to remember that this is the *formation* of a lattice from gaseous ions, so energy is released:

$Na^+(g) + Cl^-(g) \rightarrow Na^+Cl^-(s)$ $\Delta H^\ominus = -787\,kJ\,mol^{-1}$

The lattice energy depends on a number of factors since it is effectively a measurement of the strength of the (usually) ionic (electrovalent) bond between the particles. This depends on both the size of the particles and the charges they carry. The size also affects the arrangement of particles in the lattice (see Figure 3.2). Table 5.1 shows the lattice energies (L.E.) of some salts.

Table 5.1

Salt	L.E./ $kJ\,mol^{-1}$	Salt	L.E./ $kJ\,mol^{-1}$	Salt	L.E./ $kJ\,mol^{-1}$	Salt	L.E./ $kJ\,mol^{-1}$
LiF	−1036	Li*Cl*	−853	LiBr	−807	LiI	−757
NaF	−923	Na*Cl*	−787	NaBr	−747	NaI	−704
KF	−821	K*Cl*	−715	KBr	−682	KI	−649

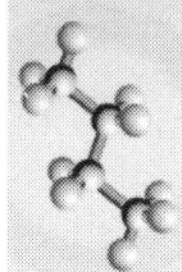

It can be seen that as the ionic radius of either ion increases, the lattice energy decreases.

The effect of changing the charge on one of the ions is shown in Table 5.2.

Table 5.2

Salt	L.E./kJ mol^{-1}	Salt	L.E./kJ mol^{-1}
NaOH	−900	Na_2O	−2481
$Mg(OH)_2$	−3006	MgO	−3791
$Al(OH)_3$	−5627	Al_2O_3	−15 916

This shows that the lattice energy increases rapidly as the charge on the ions increases.

A number of the enthalpy changes described can be measured practically, but some have to be determined indirectly from other measurements.

Thermochemical experiments are often presented in the practical paper, so it is important to understand how to use the measurements you make. The experiments generally use simple apparatus. You could be asked to determine, for example, the enthalpy of neutralisation of an acid or the enthalpy of displacement of copper by zinc.

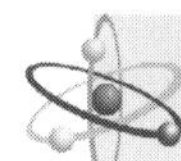

Sample practical

- A known volume of acid of known concentration is measured into a polystyrene cup.
- The temperature of the acid is measured every minute for 4 minutes.
- At the fifth minute a known volume of alkali of similar concentration is added.
- The temperature is then measured every 30 seconds for the next 3 minutes.
- A graph of temperature against time is plotted, such as the one below:

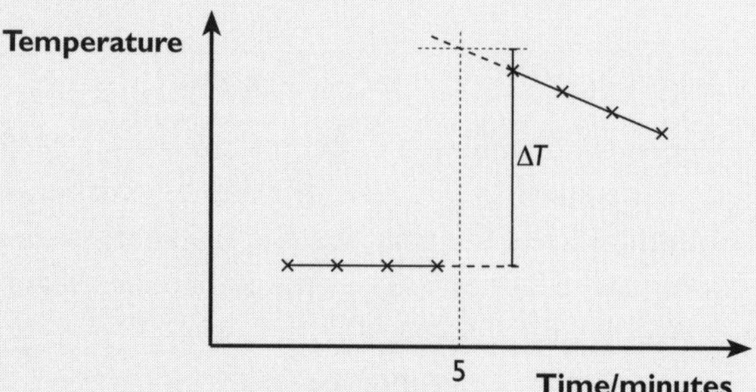

- The higher temperature line is extrapolated back to minute 5 to correct for any heat loss from the apparatus.
- The enthalpy change, ΔH, is calculated from $mc\Delta T$, where m is the mass of solution (for dilute solutions the same as its volume), c is the specific heat capacity of the solution (usually taken to be that of water) and ΔT is the temperature change based on the extrapolated value.

- The enthalpy change per mole of water formed is then calculated in kJ.

This basic method can be used for determining different enthalpy changes.

Hess's law

In simple terms, Hess's law states that the enthalpy change for the chemical process **X** → **Y** is the same whichever route is taken from **X** to **Y** provided that the states of **X** and **Y** are the same in all routes. Using **standard enthalpies** avoids this problem since the states of **X** and **Y** are defined.

This means that when it is difficult to measure enthalpy changes experimentally, other data can be used. For example, the enthalpy of reaction can be calculated using enthalpies of formation:

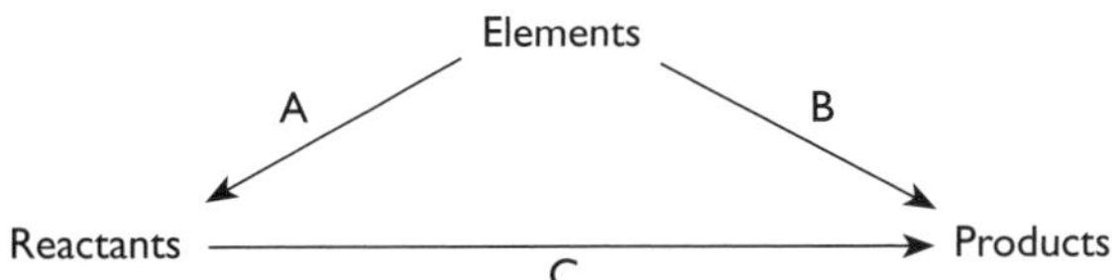

A is (sum of the enthalpies of formation of the reactants)

B is (sum of the enthalpies of formation of the products)

C is therefore (**B** – **A**)

You can see how this works in practice, with the following example.

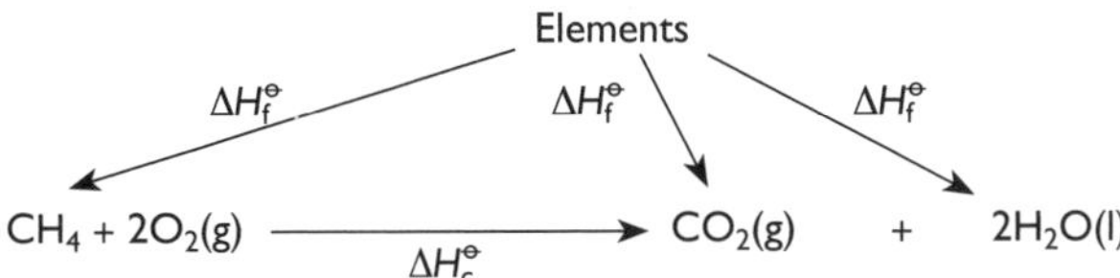

This shows how the enthalpy change for the combustion of methane, CH_4, can be found using the enthalpies of formation of methane, carbon dioxide and water.

$\Delta H_c^{\ominus}(CH_4) = \Delta H_f^{\ominus}(CO_2) + 2\,\Delta H_f^{\ominus}(H_2O) - \Delta H_f^{\ominus}(CH_4)$

Inserting the values of $\Delta H_f^{\ominus}$ for the three compounds gives:

$\Delta H_c^{\ominus}(CH_4) = -393.5 + 2(-285.8) - (-74.8)$

$= -393.5 - 571.6 + 74.8$

$= -890.3\,\text{kJ}\,\text{mol}^{-1}$

Reverse calculations to determine the enthalpy of formation of a compound using enthalpy of combustion data can be performed.

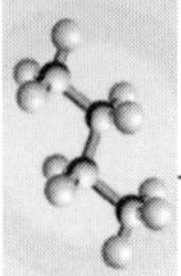

Average bond energies

Hess's law can also be used to find approximate values of $\Delta H^\ominus$ for a reaction using average bond energies (sometimes referred to as mean bond enthalpies). These average bond energies are determined from bonds in a variety of molecules. There is a list of those you are likely to need in the *Data Booklet* (also to be found at the end of the syllabus).

The Hess's law diagram for use with bond energies is shown below:

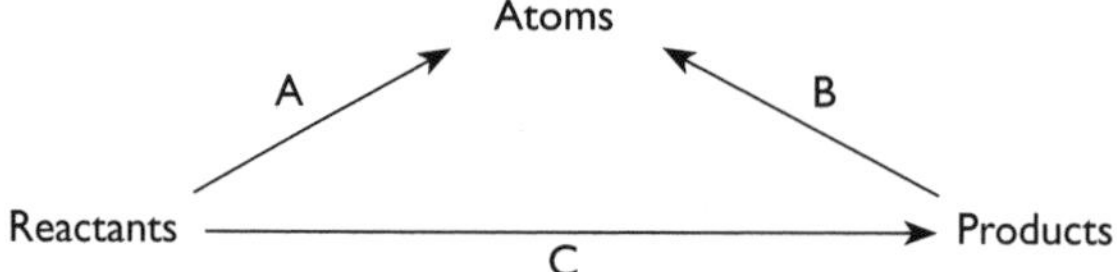

A is (sum of the bond energies of the reactants)

B is (sum of the bond energies of the products)

C is therefore (**A** – **B**)

Note that bond energies are always endothermic, i.e. they are *positive.*

When using bond energies it is useful to remember the following:

- Show the reaction using structural formulae so that you can see all the bonds present.
- Ignore bonds or groups of bonds that are unchanged in the reaction.
- Remember to change the sign of the bond energy for any bonds formed.

You need to remember something else important about reactions. Even with strongly exothermic reactions, such as the burning of magnesium, the reaction does not happen immediately magnesium is exposed to air. In order to start the reaction the magnesium has to be heated, i.e. it has to be provided with energy. This is called the **activation energy**.

Reactions can be summarised in diagrammatic form, as shown in Figure 5.2.

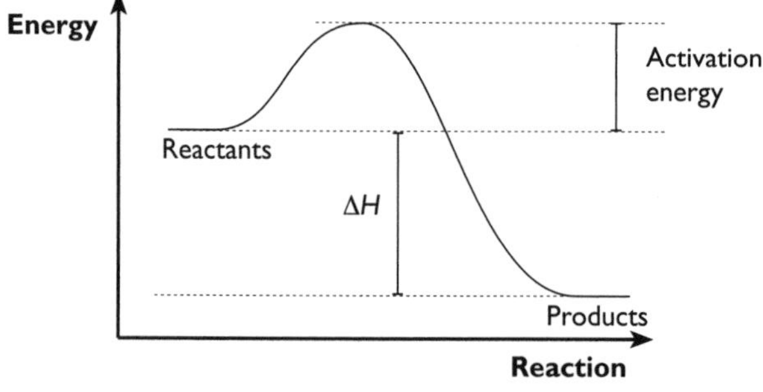

Figure 5.2 Activation energy

Some energy changes are more complicated than others. Take for example the energy of formation of an ionic (electrovalent) solid such as sodium chloride. A number of enthalpy changes are involved in converting the two elements into a solid crystal lattice. These changes can be shown diagrammatically in a **Born–Haber cycle** (Figure 5.3), which uses Hess's law.

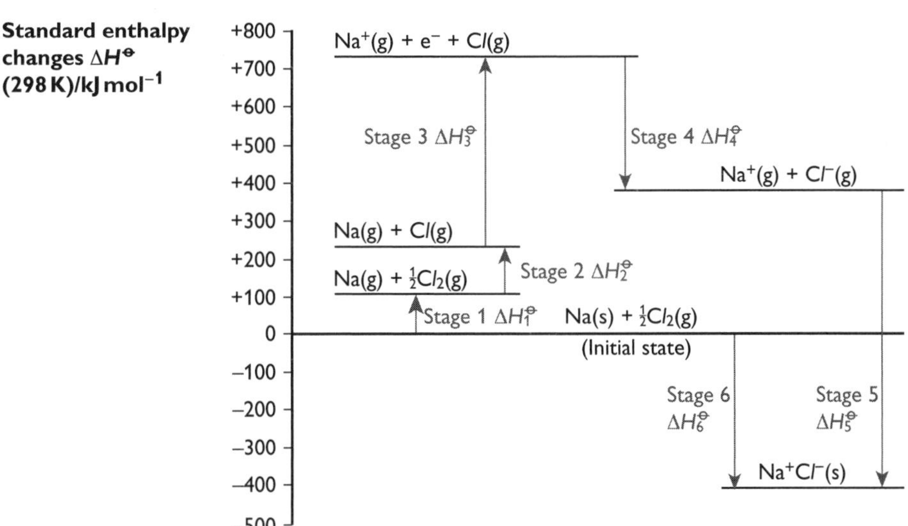

Figure 5.3 Born–Haber cycle for the enthalpy of formation of sodium chloride

Stage 1 is the enthalpy of atomisation of sodium:

$Na(s) \rightarrow Na(g)$ $\Delta H_1^\ominus = +108\,kJ\,mol^{-1}$

Stage 2 is the enthalpy of atomisation of chlorine:

$\frac{1}{2}Cl_2(g) \rightarrow Cl(g)$ $\Delta H_2^\ominus = +121\,kJ\,mol^{-1}$

Stage 3 is the ionisation energy of sodium:

$Na(g) \rightarrow Na^+(g)$ $\Delta H_3^\ominus = +496\,kJ\,mol^{-1}$

Stage 4 is the electron affinity of chlorine:

$Cl(g) \rightarrow Cl^-(g)$ $\Delta H_4^\ominus = -348\,kJ\,mol^{-1}$

Stage 5 is the lattice energy of sodium chloride (to be determined):

$Na^+(g) + Cl^-(g) \rightarrow Na^+Cl^-(s)$ $\Delta H_5^\ominus$

Stage 6 is the standard enthalpy of formation of sodium chloride:

$Na(s) + \frac{1}{2}Cl_2(g) \rightarrow Na^+Cl^-(s)$ $\Delta H_6^\ominus = -411\,kJ\,mol^{-1}$

Using Figure 5.3 you can see that:

$+108 + 121 + 496 - 348 + \Delta H_5^\ominus = -411\,kJ\,mol^{-1}$

$+377 + \Delta H_5^\ominus = -411\,kJ\,mol^{-1}$

$\Delta H_5^\ominus = -788\,kJ\,mol^{-1}$

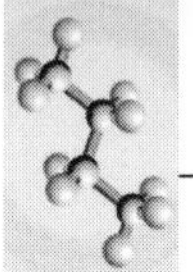

6 Electrochemistry

This chapter contains a fairly small amount of material needed for AS, and a much larger section of material which is tested at A2.

Redox processes

The word **redox** comes from two familiar words — **red**uction and **ox**idation — and refers to what happens in chemical reactions when electrons are gained or lost.

When a metal is oxidised, it loses electrons, for example:

$$Mg \rightarrow Mg^{2+} + 2e^-$$

When magnesium is burned in air, the electrons are picked up by oxygen, reducing it to the oxide ion:

$$\tfrac{1}{2}O_2 + 2e^- \rightarrow O^{2-}$$

Oxidation and reduction do not only occur with elements — for example, ions such as iron(II) ions can be oxidised to iron(III) ions:

$$Fe^{2+} \rightleftharpoons Fe^{3+} + e^-$$

Reduction of one ion to form another ion can also occur, such as in a manganate(VII) titration:

$$MnO_4^- + 8H^+ + 5e^- \rightleftharpoons Mn^{2+} + 4H_2O$$

You may be given equations and asked for the change in oxidation state or oxidation number of one of the elements present. In these cases, it is often easier to think of oxidation states/numbers on a line as shown in the diagram below:

−2 −1 0 +1 +2 +3 +4 +5 +6

Using this idea, if you place the reduced and oxidised species at the correct points on the line, you can see the change in oxidation number, or the number of electrons lost or gained.

Consider the oxidation of hydrogen sulfide to form sulfur dioxide:

$$2H_2S(g) + 3O_2(g) \rightarrow 2SO_2(g) + 2H_2O(l)$$

What is the change in oxidation number of sulfur?

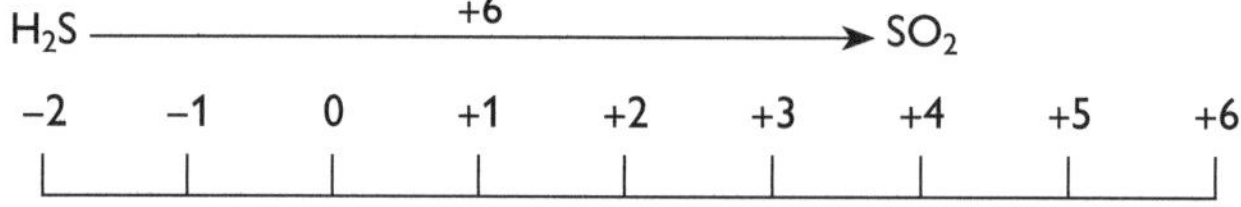

As you can see from the diagram, you simply measure the distance between the two oxidation states.

In working out oxidation numbers there are some simple rules:

- Atoms of uncombined elements have an oxidation number of zero.
- Simple ions have an oxidation number equal to the charge on the ion.
- In complex ions, the sum of the oxidation numbers in all the elements present equals the overall charge on the ion.

Try this yourself

(20) State the oxidation number of iron in each of the following substances:

(a) Fe_2O_3

(b) Fe

(c) Na_2FeO_4

(21) What are the oxidation states of manganese in the following equation?

$$2KMnO_4 + MnO_2 + 4KOH \rightarrow 3K_2MnO_4 + 2H_2O$$

Electrochemistry in industry

The use of electricity to bring about chemical change is an important part of the chemical industry and you need to know about three processes for AS.

The diaphragm cell

Chlorine has been produced by electrolysis for many years, and the most recent method uses the diaphragm cell shown in Figure 6.1.

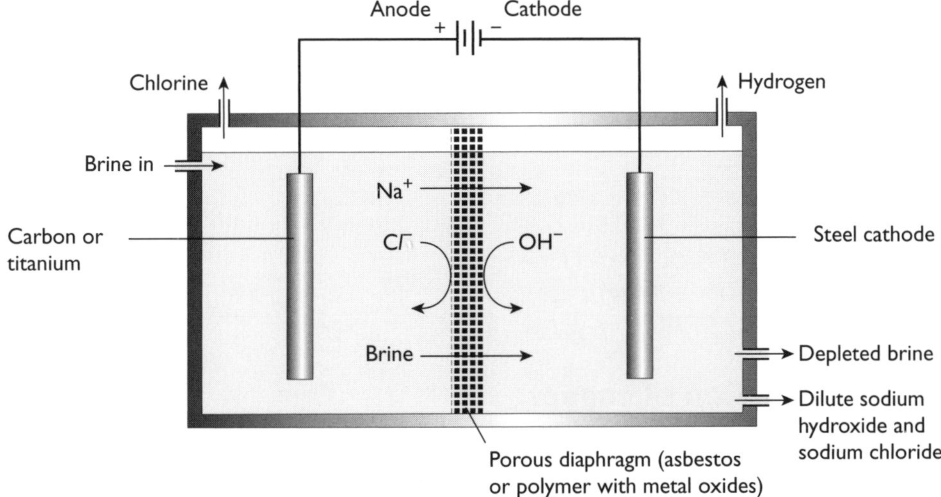

Figure 6.1 A diaphragm cell

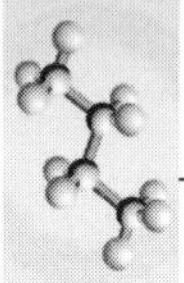

The key factors to remember are that:

- the electrodes are separated by a diaphragm, usually made of asbestos
- the electrolyte is brine (a solution of sodium chloride)
- the anode is usually made of titanium
- the cathode is usually made of steel
- chlorine gas is produced at the anode, hydrogen gas at the cathode:
  anode: $2Cl^-(aq) \rightarrow Cl_2(g) + 2e^-$
  cathode: $2H_2O(l) + 2e^- \rightarrow H_2(g) + 2OH^-(aq)$
- sodium hydroxide is an important by-product

The extraction of aluminium

Aluminium is too reactive a metal to be extracted by heating with coke (carbon), like other metals such as iron. It is extracted by electrolysing aluminium oxide dissolved in molten cryolite, Na_3AlF_6 (Figure 6.2).

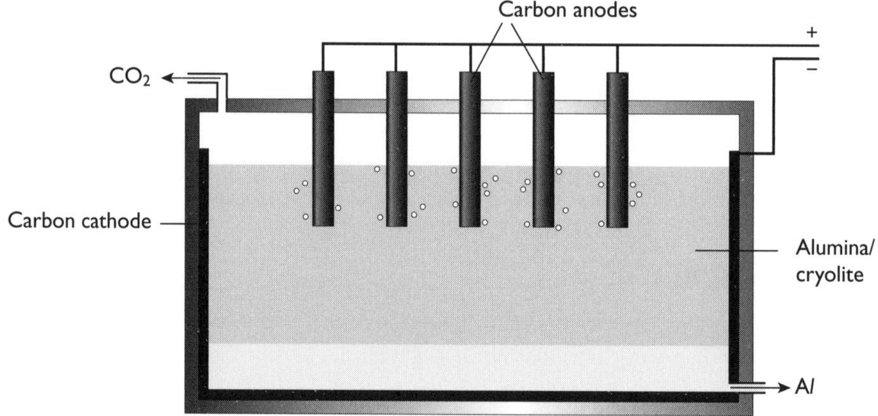

Figure 6.2 Extraction of aluminium

The key factors to remember are that:

- the electrolyte is aluminium oxide dissolved in molten cryolite
- both the anode and cathode are made of carbon, but the anodes have to be replaced regularly because they are oxidised
- oxygen gas is produced at the anode, aluminium at the cathode (note that at the temperature of the cell the aluminium is molten):
  anode: $6O^{2-} \rightarrow 3O_2(g) + 12e^-$
  cathode: $4Al^{3+} + 12e^- \rightarrow 4Al\,(l)$

The purification of copper

Copper is not *extracted* using electrolysis, it is *purified* using this process (Figure 6.3). This is because large amounts of copper are used in electrical wiring and the presence of impurities would increase the resistance of the wire, resulting in overheating.

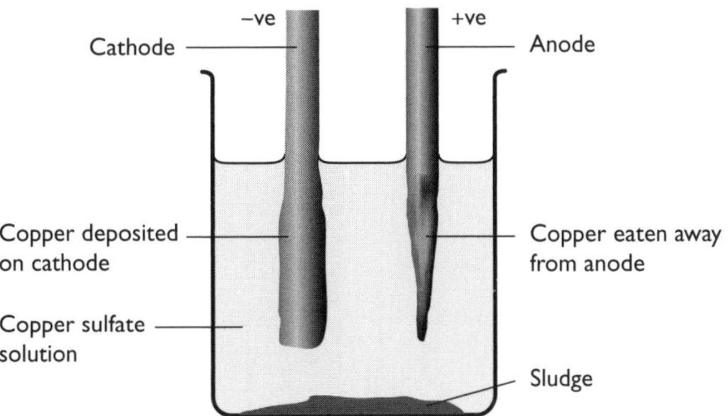

Figure 6.3 Purification of copper

The key factors to remember are that:

- the electrolyte is an aqueous solution of copper(II) sulfate
- the cathode is pure copper; the anode is impure copper
- copper dissolves from the anode and is deposited on the cathode:
  anode: $Cu(s) \rightarrow Cu^{2+}(aq) + 2e^-$
  cathode: $Cu^{2+}(aq) + 2e^- \rightarrow Cu(s)$
- impurities such as silver are deposited on the bottom of the cell as a sludge; they are recovered to help reduce the cost of the process

Electrode potentials

There are two aspects to electrochemistry; the first, electrolysis, involves using electricity to bring about chemical reactions. The second, electrode potentials, relies on chemical reactions to generate an electric current.

It is important that you learn two definitions linked to electrode potentials.

Standard electrode potentials

A **standard electrode (redox) potential** is defined as the electrode potential measured under standard conditions (temperature 298 K, 1 atmosphere pressure, 1 mole of the redox participants of the half-reaction) against a standard hydrogen electrode.

A standard hydrogen electrode is shown in Figure 6.4.

Tip It is important that you can draw a standard hydrogen electrode in an exam.

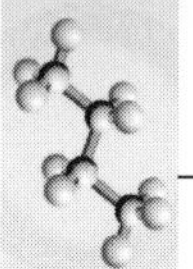

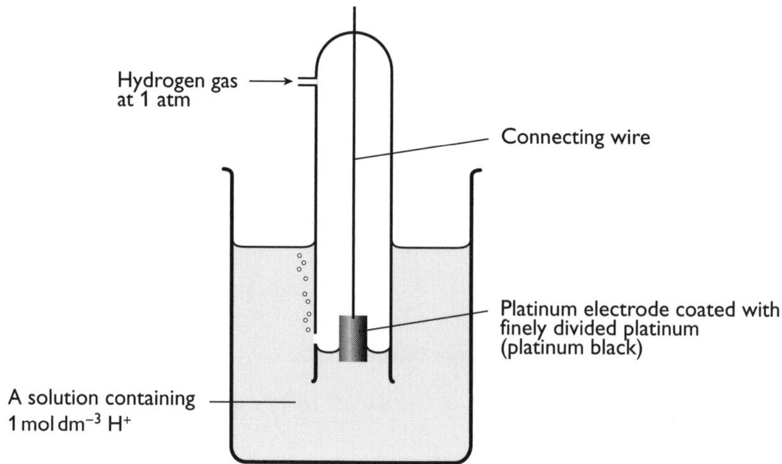

Figure 6.4 A standard hydrogen electrode

In Figure 6.5 you can see how the standard electrode potential of another electrode can be measured using the hydrogen electrode.

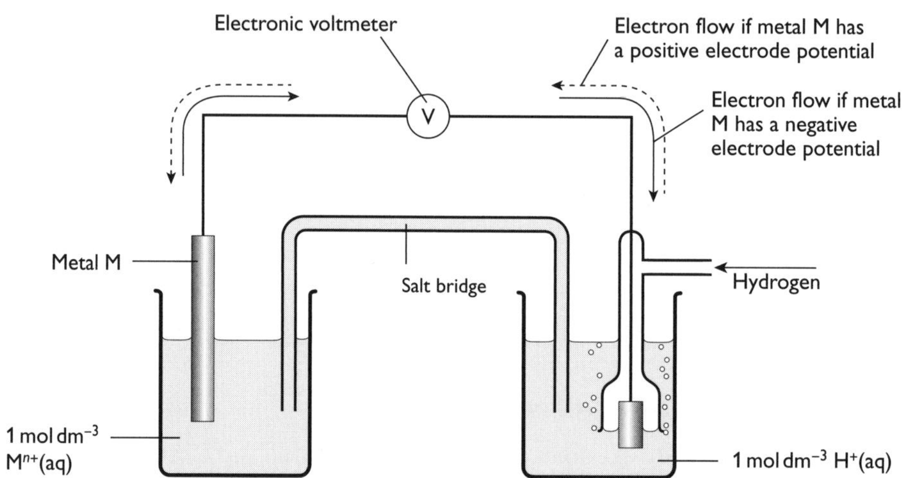

Figure 6.5 Measuring a standard electrode potential

Standard cell potentials

A **standard cell potential** is the potential produced when two standard electrodes are connected to form a cell such as that in Figure 6.6.

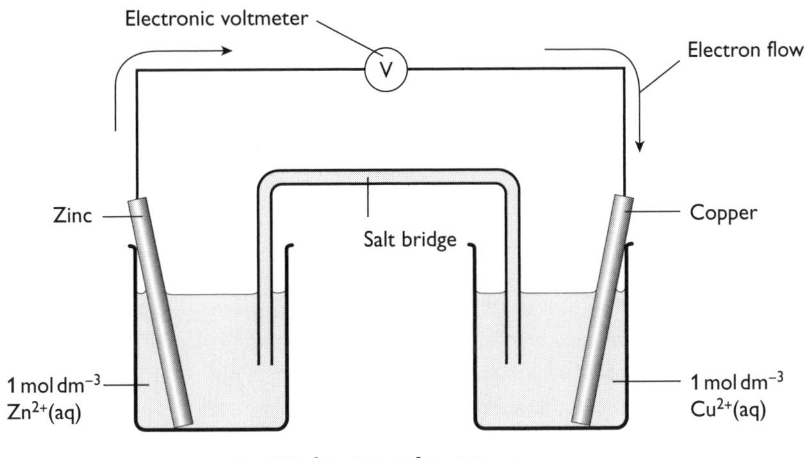

Figure 6.6 Apparatus for determining the standard cell potential

The cell potential has a contribution from the anode, which is a measure of its ability to lose electrons — its **oxidation potential**. The cathode has a contribution based on its ability to gain electrons — its **reduction potential.** The cell potential can then be written as:

E_{cell} = oxidation potential + reduction potential

As with oxidation numbers, to calculate the overall cell potential it helps to work using a linear scale from negative to positive:

−0.8	−0.6	−0.4	−0.2	0	+0.2	+0.4	+0.6	+0.8

In the cell shown in Figure 6.6, the oxidation potential of the zinc anode is +0.76 volts, and the reduction potential of the copper cathode is +0.34 volts.

$E_{cell} = 0.76 + 0.34$ volts = 1.10 volts

Care must be taken to change the signs given in electrode potential tables to reflect what is happening. These data are given in terms of the *reduction* of the ions concerned. When you are showing a cell, remember to show the reactants with oxidation occurring on the right and reduction on the left. This means that in the cell described above we would write:

$Zn(s)/Zn^{2+}(aq)$ // $Cu^{2+}(aq)/Cu(s)$

because the zinc metal is oxidised and the copper ions are reduced.

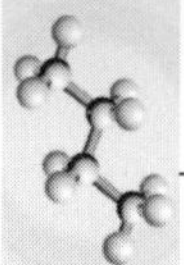

Try this yourself

(22) Use the table of standard electrode potentials in the *Data Booklet* in the syllabus to calculate the cell potentials for the following electrode pairs:

(a) $Zn(s)/Zn^{2+}(aq)$ // $Ag^{+}(aq)/Ag(s)$

(b) $Mg(s)/Mg^{2+}(aq)$ // $Pb^{2+}(aq)/Pb(s)$

(c) $Cu(s)/Cu^{2+}(aq)$ // $Ag^{+}(aq)/Ag(s)$

Non-metallic electrodes

Measuring the standard electrode potential of a non-metallic element presents different problems from using metals. However, we have already seen one way of overcoming this in the hydrogen electrode. The cell shown in Figure 6.7 shows how two non-metallic elements can form a cell.

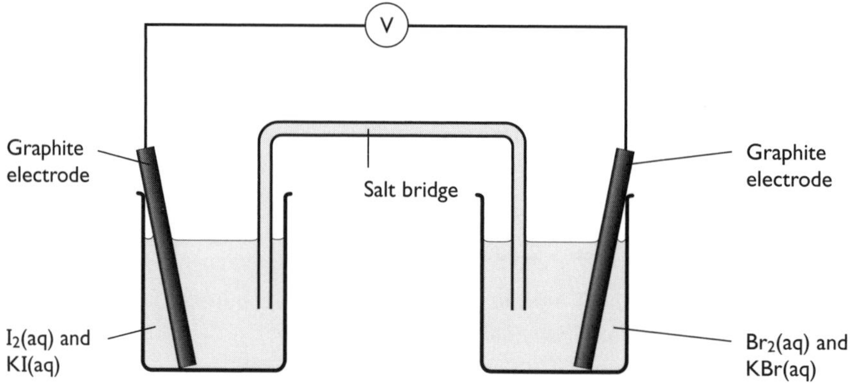

Figure 6.7 A cell formed using two non-metallic elements

The same basic technique can be used for ions of the same element in different oxidation states. The set-up can be seen in Figure 6.8. The electrodes chosen in each case are platinum, although in the laboratory you may have used carbon.

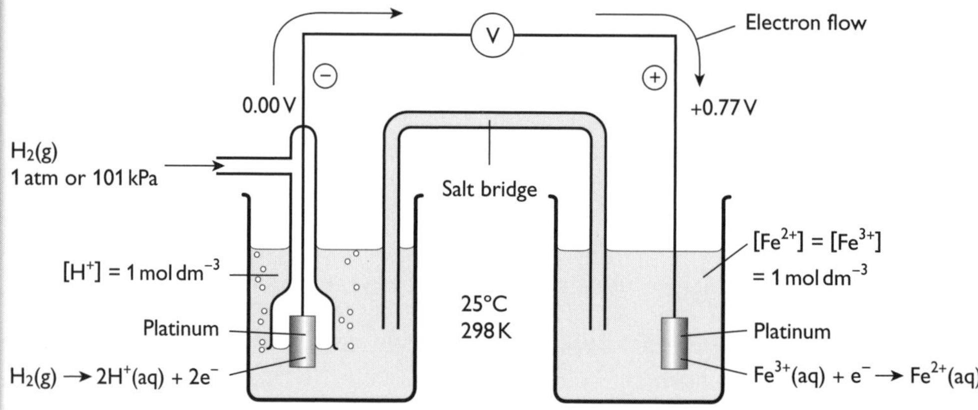

Figure 6.8 A cell formed using ions of the same element

Try this yourself

(23) Now use the table of standard electrode potentials in the *Data Booklet* in the syllabus to calculate the cell potentials for the following electrode pairs which include non-metals:

(a) $Mg(s)/Mg^{2+}(aq)$ // $\frac{1}{2}Cl_2(g)/Cl^-(aq)$
(b) $Pb(s)/Pb^{2+}(aq)$ // $\frac{1}{2}Br_2(l)/Br^-(aq)$
(c) $Fe^{2+}(aq)/Fe^{3+}(aq)$ // $\frac{1}{2}Cl_2(g)/Cl^-(aq)$
(d) $Br^-(aq)/\frac{1}{2}Br_2(l)$ // $\frac{1}{2}Cl_2(g)/Cl^-(aq)$

Predicting reactions

As well as using electrode potentials to calculate the voltage a particular combination of electrodes will produce under standard conditions, electrode potentials can also be used to predict how likely a given chemical reaction is to occur.

Table 6.1

Metal electrodes	Voltage/V	Non-metal electrodes	Voltage/V
$K^+ + e^- \rightleftharpoons K$	−2.92	$\frac{1}{2}F_2 + e^- \rightleftharpoons F^-$	+2.87
$Mg^{2+} + 2e^- \rightleftharpoons Mg$	−2.38	$MnO_4^- + 8H^+ + 5e^- \rightleftharpoons Mn^{2+} + 4H_2O$	+1.52
$Al^{3+} + 3e^- \rightleftharpoons Al$	−1.66	$\frac{1}{2}Cl_2 + e^- \rightleftharpoons Cl^-$	+1.36
$Zn^{2+} + 2e^- \rightleftharpoons Zn$	−0.76	$NO_3^- + 2H^+ + e^- \rightleftharpoons NO_2 + H_2O$	+0.81

In Table 6.1, the half-reactions with a high *negative* E° do *not* take place readily (think of the reactions of the alkali (Group I) metals with water). By contrast, those with a high *positive* E° take place spontaneously (think of manganate(VII) as an oxidising agent).

The half-equations can be used to construct full chemical equations, by making sure that the numbers of electrons balance and that the overall cell potential is positive. There are some key points that help you to get this right:

- Write the half-equation with the more negative E° first.
- Remember that the more positive E° will be a reduction.
- Draw anti-clockwise arrows to help predict the overall reaction.

For example, consider the reaction between Fe^{2+} ions and chlorine gas. The two half-equations are:

$Fe^{3+} + e^- \rightleftharpoons Fe^{2+}$ $\quad E^\circ = +0.77\,V$

$\frac{1}{2}Cl_2 + e^- \rightleftharpoons Cl^-$ $\quad E^\circ = +1.36\,V$

Reversing the first equation and adding gives:

$Fe^{2+} + \frac{1}{2}Cl_2 \rightarrow Fe^{3+} + Cl^-$ $\quad E^\circ_{cell} = +0.59\,V$

E°_{cell} is positive. Therefore, when chlorine gas oxidises iron(II) ions to iron(III) ions, it is reduced to chloride ions.

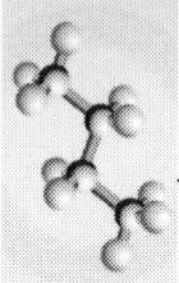

The effects of non-standard conditions

For most of the work you will do with electrode potentials, you can assume that the conditions are standard, but you do need to know the effects of changing the concentration of the aqueous ion.

Think of the half-cell reaction:

$M^{n+}(aq) + ne^- \rightleftharpoons M$

Le Chatelier's principle (Chapter 7) predicts that if the concentration of $M^{n+}(aq)$ is increased, then the equilibrium should move to the right. If this were the case, the electrode should become more positive with respect to the solution. Hence, the electrode potential should also become more positive. This also means that the half-cell would become a better oxidising agent. Reducing the concentration of $M^{n+}(aq)$ would have the opposite effect.

Batteries and fuel cells

'Battery' is the name commonly given to portable sources of electric current and it is often inaccurate. Many so-called 'batteries' are simple cells, of which the most common is the zinc/alkali or 'dry' cell (Figure 6.9). Such cells are portable and much more practical than 'wet' cells containing aqueous solutions. Others, such as 9 V batteries or 12 V car batteries, are indeed 'batteries of cells'.

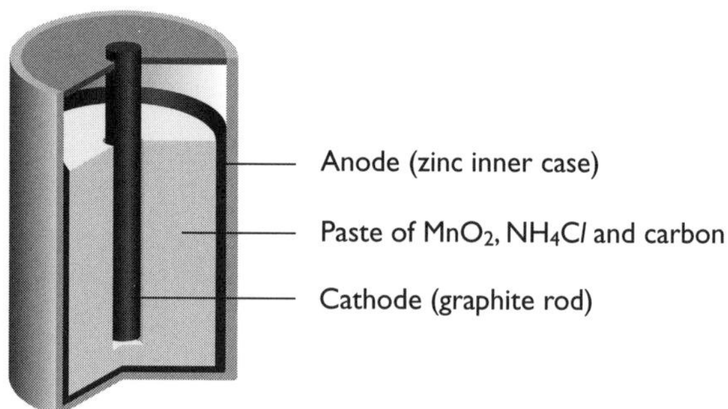

Figure 6.9 A dry cell

The reactions in a dry cell are as follows:

$Zn(s) \rightarrow Zn^{2+}(aq) + 2e^-$

$MnO_2(s) + H_2O(l) + e^- \rightarrow MnO(OH) + OH^-(aq)$

The problem with all such cells is that they have a finite lifetime because the materials get used up.

Batteries of various sorts have been developed over the years giving different voltages, longer life and the ability to be recharged. This is a result of the huge increase in the use of portable electronic devices such as iPods, computers and mobile phones.

There has also been considerable interest in the development of electrically powered cars. Conventional batteries of the lead/acid type are far too heavy and bulky to provide the power needed, so there has been research into different ways of supplying this power. One possible development, which allows a continuous supply of electricity, is the **fuel cell.** This consists of a fuel (usually hydrogen gas) and an oxidant (oxygen gas) being pumped through the cell, shown in Figure 6.10.

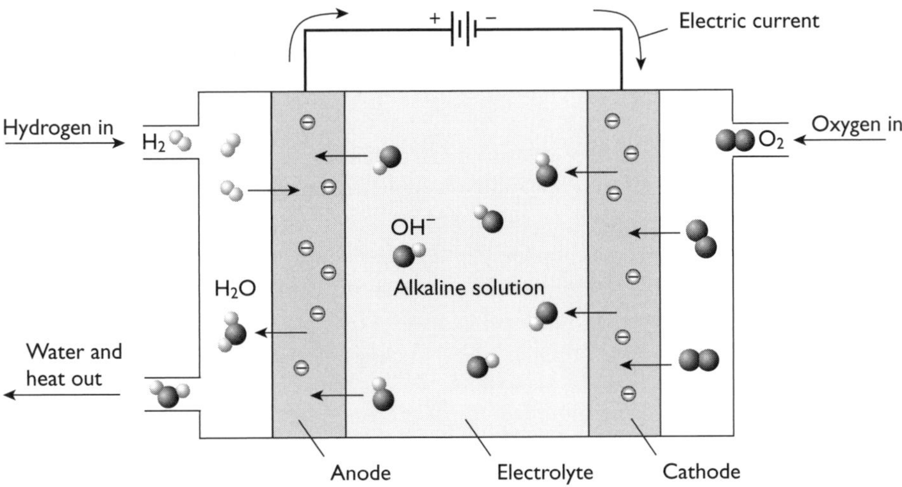

Figure 6.10 A fuel cell

The technology is relatively new and the challenge is to make sure the cells are efficient and as small and light as possible. One disadvantage of the cell shown in Figure 6.10 is that it is easily poisoned by carbon dioxide, which reacts with the alkali.

Electrolysis

The converse of using chemical reactions to generate electricity is to use electricity to bring about chemical reactions. You have already seen examples of this in the chemical industry, at the start of this section.

Like all chemical reactions, it is possible to calculate quantities of materials used or produced in such a process. In the case of electrolysis we know, for example, how many electrons are needed to produce one atom of aluminium:

$$Al^{3+} + 3e^- \rightarrow Al$$

The quantity of electricity needed to produce 1 mole of aluminium depends on the number of electrons needed, the charge on the electron and the number of atoms. The magnitude of the charge per mole of electrons is known as the Faraday constant and is generally quoted as 96 500 C mol^{-1}. This is expressed as:

$$F = Le$$

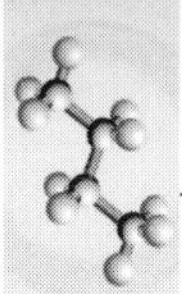

where L is the Avogadro constant and e the charge on the electron.

In an examination you might be asked to calculate the quantity of charge passed during electrolysis, or the mass and/or volume of a substance liberated during the process. Let's look at how you might tackle this.

Example

What mass of copper is produced at the cathode when 2.40 A are passed through a solution of copper(II) sulfate for 25 minutes?

Answer

First, calculate the number of coulombs of charge passed. Remember that charge equals current in amps multiplied by time in seconds:

$2.40 \times 25 \times 60 = 3600\,\text{C}$

Next, calculate how many moles of electrons this corresponds to:

$\frac{3600}{96\,500} = 3.73 \times 10^{-2}\,\text{mol}$

Since copper(II) ions have a charge of +2 this corresponds to:

$\frac{3.73 \times 10^{-2}}{2}$ mol of copper

Therefore the mass of copper produced is $\frac{3.73 \times 10^{-2}}{2} \times 63.5 = 1.18\,\text{g}$

For the electrolysis of a solution of sodium sulfate, Na_2SO_4, or sulfuric acid, H_2SO_4, the product at the cathode is hydrogen gas. Instead of using mass with hydrogen, we can use the volume as a fraction of the molar volume of a gas, $24\,\text{dm}^3$.

A value for the Avogadro constant can be determined using electrolysis. This method requires the measurement of the mass of an element, such as copper, produced in a fixed period of time and at a constant known current, to be measured as accurately as possible. Knowing the charge on the copper(II) ion we can then use $F = Le$ to determine L, the Avogadro constant.

Some exam questions may ask you to predict what substances are liberated at a given electrode. To answer this you need to check if the electrolysis is taking place in the molten salt or in aqueous solution, and consider the electrode potentials and concentrations of the ions.

- If it is a molten salt there can only be one element discharged at each electrode.
- If it is an aqueous solution it is possible for hydrogen to be discharged at the cathode and oxygen at the anode:

  $4H^+ + 4e^- \rightarrow 2H_2(g)$ $\qquad E^\ominus = 0.00$

  $4OH^- - 4e^- \rightarrow 2H_2O(l) + O_2(g)$ $\qquad E^\ominus = -0.40$

  In order to decide which ions are discharged, you need to compare $E^\ominus$ for the half-cell reactions above with those of the other ions in solution.

Think about the electrolysis of a concentrated solution of sodium chloride. Hydrogen is produced at the cathode because hydrogen ions accept electrons more easily than sodium ions. On the other hand, chlorine is produced at the anode because chloride ions are present in a higher concentration than hydroxide ions.

7 Equilibria

This first part of this chapter deals with material needed for the AS exam. The second part is only needed for the A2 exam.

Factors affecting chemical equilibria

You probably think you know what the word 'equilibrium' means, and you may have used it in physics to mean 'forces in balance' or something similar. In chemistry it has a similar meaning, but here it is chemical reactions that are in balance. The sign $\rightleftharpoons$ is used to represent a reaction in equilibrium. What exactly does the word mean in a chemical context?

First, you need to understand that all chemical reactions are reversible, given enough energy. In a reversible reaction at equilibrium, the rates of the forward and reverse reactions are the same. In other words, the amount of reactants forming products in a given time is the same as the amount of products breaking down to give reactants in the same time. These are generally referred to as **dynamic equilibria**.

An example of a dynamic equilibrium in the gas phase is the reaction of nitrogen and hydrogen to form ammonia in the Haber process:

$$N_2(g) + 3H_2(g) \rightleftharpoons 2NH_3(g)$$

Dynamic equilibria can also take place in the liquid phase:

$$2H^+(aq) + 2CrO_4^{2-}(aq) \rightleftharpoons Cr_2O_7^{2-}(aq) + H_2O(l)$$

In a dynamic equilibrium:

- the forward reaction takes place at the same rate as the reverse reaction
- there is no net change in the concentration of each substance
- the equilibrium compositions of the substances can be approached from either reactants or products

Le Chatelier's principle

The French chemist Henri Le Chatelier studied many dynamic equilibria and suggested a general rule to help predict the change in the position of equilibrium. You should learn the rule, but think carefully when you apply it in exam questions.

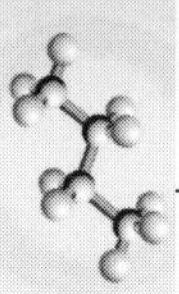

Le Chatelier's principle states that if a closed system at equilibrium is subject to a change, then the system will adjust in such a way as to minimise the effect of the change.

The factors that can be changed easily are concentration, temperature and pressure. You might also have suggested 'using a catalyst' — this possibility is looked at later.

Consider the following general reaction:

$m\text{A} + n\text{B} \rightleftharpoons p\text{C} + q\text{D}$ $\qquad\qquad$ ΔH is negative (exothermic)

If we *increase* the concentration of *either* of the reactants A or B, more of the products C and D will be produced.

The general reaction shown above is an exothermic reaction. If we *increase* the temperature, there will be less of the products formed. The reverse would be true for an endothermic reaction.

The influence of pressure is only relevant for gas phase reactions. Think back to the Haber process for making ammonia:

$N_2(g) + 3H_2(g) \rightleftharpoons 2NH_3(g)$

There are fewer molecules on the right-hand side of the equation. This means that if we *increase* the pressure, the equilibrium will shift to produce more ammonia, reducing the total number of molecules in the system and therefore reducing the pressure.

Try this yourself

(24) For each of the following reactions, use Le Chatelier's principle to decide what will be the effect on the position of equilibrium as a result of the change stated:

- **(a)** $2NO(g) + O_2(g) \rightleftharpoons 2NO_2(g)$ ΔH is negative
  - **(i)** increase the temperature
  - **(ii)** increase the pressure
- **(b)** $C_2H_5OH(aq) + CH_3CO_2H(aq) \rightleftharpoons CH_3CO_2C_2H_5(aq) + H_2O(l)$
  - **(i)** increase the concentration of CH_3CO_2H
  - **(ii)** remove H_2O

In order to see what effect adding a catalyst might have, we have to think about the processes taking place. Look at Figure 7.1, which shows the energy profile of an equilibrium reaction with and without a catalyst present.

Figure 7.1 shows that the catalyst has lowered the activation energy for the forward reaction. Consider the reverse reaction — the activation energy for this has also been lowered by the same amount. In other words, the presence of a catalyst does *not* change the position of equilibrium; it enables equilibrium to be established more quickly.

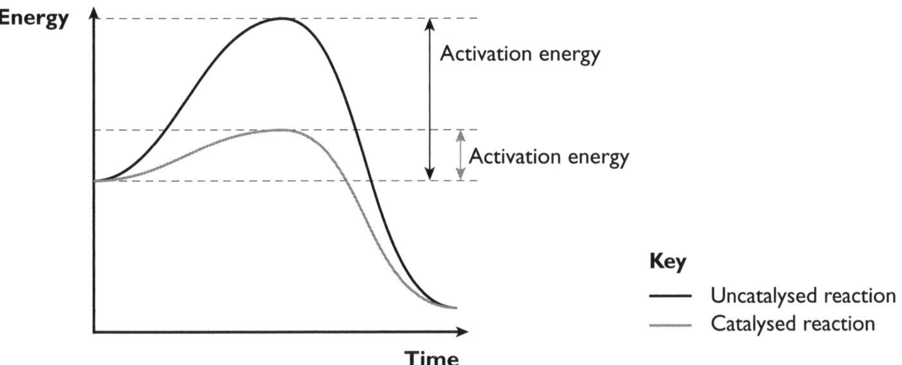

Figure 7.1 Reaction profile with and without a catalyst

Equilibrium constants and calculations

For any equilibrium in the liquid state, an equilibrium constant, K_c, can be defined in terms of concentration. For our general equation:

$$mA + nB \rightleftharpoons pC + qD$$

$$K_c = \frac{[C]^p[D]^q}{[A]^m[B]^n}$$

The quantities in square brackets represent the concentrations of the different species at equilibrium. It is important to remember that equilibrium constants depend on temperature.

For gas phase reactions, partial pressures (in atmospheres) of the species involved in the equilibrium are used. To distinguish this from reactions in solution the symbol K_p is used for the equilibrium constant. So, for the reaction between hydrogen and iodine to form hydrogen iodide:

$$H_2(g) + I_2(g) \rightleftharpoons 2HI(g)$$

$$K_p = \frac{p(HI)p(HI)}{p(H_2)p(I_2)} = \frac{p(HI)^2}{p(H_2)p(I_2)}$$

where p represents the partial pressure of each of the species

Let's look at how these two constants are calculated.

Consider the reaction between ethanol and ethanoic acid:

$$C_2H_5OH(aq) + CH_3CO_2H(aq) \rightleftharpoons CH_3CO_2C_2H_5(aq) + H_2O(l)$$

In an experiment, 0.100 mol of ethanol and 0.200 mol of ethanoic acid were mixed together and the mixture was allowed to reach equilibrium. The acid was then titrated with 1.00 mol dm^{-3} sodium hydroxide and 115 cm^3 were needed to neutralise the acid. This volume of sodium hydroxide contains:

$$\frac{1.00 \times 115}{1000} = 0.115 \text{ mol of sodium hydroxide}$$

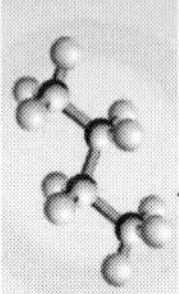

This means that there were 0.115 mol of ethanoic acid present at equilibrium.

Now let's look at the other species present.

	$C_2H_5OH(aq)$	$CH_3CO_2H(aq)$	$CH_3CO_2C_2H_5(aq)$	$H_2O(l)$
Start moles	0.100 mol/V	0.200 mol/V	0.0 mol/V	0.0 mol/V
Equilibrium moles	? mol/V	0.115 mol/V	? mol/V	? mol/V

We have to divide by the total volume, V, since we are using concentrations.

We can deduce that (0.200 – 0.115) mol of ethanoic acid have been used. Hence the same amount of ethanol will have reacted. So, at equilibrium, the moles present are:

	$C_2H_5OH(aq)$	$CH_3CO_2H(aq)$	$CH_3CO_2C_2H_5(aq)$	$H_2O(l)$
Equilibrium moles	0.015 mol/V	0.115 mol/V	0.085 mol/V	0.085 mol/V

Substituting into the expression for K_c, since there are the same number of molecules on each side of the equilibrium, the V terms cancel out:

$$K_c = \frac{0.085 \times 0.085}{0.015 \times 0.115} = 4.19$$

We can use a similar process to calculate K_p for reactions taking place in the gas phase. To do this we need to understand what is meant by **partial pressure**. The partial pressure of a gas is defined as its mole fraction multiplied by the total pressure. Suppose we think of air as consisting of one-fifth oxygen and four-fifths nitrogen at a total pressure of 100 kPa. Then:

partial pressure of oxygen, $p(O_2) = 1/5 \times 100\,\text{kPa} = 20\,\text{kPa}$

partial pressure of nitrogen, $p(N_2) = 4/5 \times 100\,\text{kPa} = 80\,\text{kPa}$

Together this gives the total pressure of 100 kPa.

Let's consider the dissociation of hydrogen iodide at 700 K:

$$2HI(g) \rightleftharpoons H_2(g) + I_2(g)$$

If the value of K_p under these conditions is 0.020, and the reaction started with pure hydrogen iodide at a pressure of 100 kPa, what will be the partial pressure of hydrogen, x, at equilibrium?

	2HI(g)	$H_2(g)$	$I_2(g)$
Start moles	100 kPa	0 kPa	0 kPa
Equilibrium moles	(100 – 2x) kPa	x kPa	x kPa

$$K_p = \frac{p(H_2) \times p(I_2)}{p(HI)^2}$$

Substituting in the expression for K_p gives:

$$0.020 = \frac{x \times x}{(100 - 2x)^2} = \frac{x^2}{(100 - 2x)^2}$$

Taking the square root of each side gives: $0.141 = \frac{x}{(100 - 2x)}$

Rearranging gives: $14.1 - 0.282x = x$

Or : $14.1 = 1.282x$

Hence: $x = 11.0$ kPa (to 3 s.f.)

The Haber process

In the Haber (or Haber–Bosch) process, ammonia is produced on a massive scale using nitrogen from the air and hydrogen from the reaction of methane, CH_4, with steam:

$$N_2(g) + 3H_2(g) \rightleftharpoons 2NH_3(g) \qquad \Delta H^\circ = -92\,kJ\,mol^{-1}$$

To make the process as economic as possible, the following are needed:

- a high equilibrium concentration of ammonia
- equilibrium to be reached in a short time

In order to understand the conditions chosen you need to know more about the reaction.

First, the equilibrium constant at 298 K is very high, but only a tiny amount of ammonia is produced at this temperature because the rates of the forward and reverse reactions are so low that equilibrium is never reached.

Second, increasing the temperature increases the rates of reaction but drastically reduces the equilibrium constant because the reaction is exothermic.

Third, increasing the partial pressures of the reactants increases the equilibrium concentration of ammonia — for example, using 7500 kPa of hydrogen and 2500 kPa of nitrogen at 798 K gives about 10% of ammonia at equilibrium.

The solution is to use an iron catalyst to increase the rate of attaining equilibrium, and then to use a compromise of the conditions — a relatively low temperature (around 750 K) and a moderately high pressure (20 000 kPa). The equilibrium mixture is then passed through a heat exchanger to cool and liquefy the ammonia, which is removed, and the unreacted nitrogen and hydrogen are recycled.

The Contact process

The key stage in the production of sulfuric acid also relies on an equilibrium reaction. In this process, sulfur dioxide is reacted with oxygen to form sulfur trioxide:

$$2SO_2(g) + O_2(g) \rightleftharpoons 2SO_3(g) \ \Delta H^\circ = -197\,kJ\,mol^{-1}$$

The forward reaction is exothermic. Therefore, Le Chatelier's principle predicts that cooling the reaction would produce the maximum yield. However, as you saw in the Haber process, this reduces the rate of attaining equilibrium.

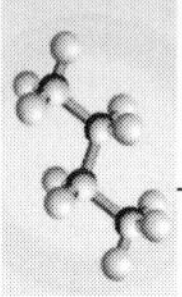

Since there is an overall reduction in the number of molecules moving from left to right, Le Chatelier's principle also predicts that increasing the pressure will drive the equilibrium to the right, increasing the yield of sulfur trioxide.

The conditions used mirror these principles. The gas mixture is passed over three catalyst beds and is cooled after each passage to try to force the equilibrium to the right. Although higher pressure is predicted to push the equilibrium to the right, most chemical plants producing sulfuric acid operate at just above atmospheric pressure to reduce costs.

Ionic equilibria

Bronsted–Lowry acids and bases

The Bronsted–Lowry theory is the most commonly used description of acidity. It describes an acid as a substance capable of donating protons (H^+) and a base as a substance capable of accepting them:

$$HCl + NH_3 \rightarrow Cl^- + NH_4^+$$

Here hydrogen chloride is acting as an acid by donating a proton to ammonia, which is acting as a base. In the reverse of this reaction, the chloride ion can accept a proton from the ammonium ion. This means that the ammonium ion would be acting as an acid in donating the proton to the chloride ion, which would be acting as a base. In this reaction, the chloride ion is known as the **conjugate base** of hydrogen chloride and the ammonium ion as the **conjugate acid** of ammonia.

It is also important to know how substances behave with water, the common solvent. Hydrogen chloride reacts with water as follows:

$$HCl + H_2O \rightarrow Cl^- + H_3O^+$$

Here water is acting as a base, accepting a proton from hydrogen chloride; the H_3O^+ ion is its conjugate acid.

Water also reacts with ammonia:

$$H_2O + NH_3 \rightarrow OH^- + NH_4^+$$

Water is acting as an acid, donating a proton to ammonia; the OH^- ion is its conjugate base.

Try this yourself

(25) Look at the following equation and label the conjugate acid–base pairs.

$$H_2CO_3 + H_2O \rightarrow HCO_3^- + H_3O^+$$

Strong and weak acids and bases

It is really important to understand the difference between *dilute* and *weak*, and *concentrated* and *strong* when you are talking about acids and bases:

- A **strong acid** or **strong base** is *completely* ionised in solution.
- A **weak acid** or **weak base** is only *partially* ionised in solution.
- A **concentrated acid** or **concentrated base** has a *large* number of moles per unit volume of the acid or base concerned.
- A **dilute** solution has a *small* number of moles relative to the solvent.

Examples of strong and weak acids and bases are shown in Table 7.1.

Table 7.1

Acids				Bases			
Strong	**pH**	**Weak**	**pH**	**Strong**	**pH**	**Weak**	**pH**
HC*l*	1	CH_3CO_2H	3	NaOH	14	NH_3	11
HNO_3	1	H_2CO_3	4	KOH	14	CH_3NH_2	12
H_2SO_4	1	H_2O	7	$Ca(OH)_2$	12	H_2O	7

pH values are given to the nearest whole number for 0.1 mol dm^{-3} solutions.

The categories 'strong' and 'weak' are qualitative. To be accurate we need to be able to make quantitative comparisons using dissociation constants or the pH of solutions of the compounds.

When an acid HA dissolves in water, the following equilibrium is established:

$$HA(aq) + H_2O(l) \rightleftharpoons H^+(aq) + A^-(aq)$$

The position of the equilibrium indicates the strength of the acid. For a strong acid, the equilibrium favours the products, and the reaction goes almost to completion. For a weak acid the equilibrium favours the reactants with a relatively small amount of $H^+(aq)$ produced.

A more precise indication of the position of equilibrium can be obtained by using an equilibrium constant:

$$K = \frac{[H^+(aq)]\,[A^-(aq)]}{[HA(aq)]\,[H_2O(l)]}$$

In fact the $[H_2O(l)]$ is almost constant, so we use a new equilibrium constant, K_a, called the **acid dissociation constant**, which includes this and has units of mol dm^{-3}.

$$K_a = \frac{[H^+(aq)]\,[A^-(aq)]}{[HA(aq)]}$$

Values of K_a are small, and it is usual to convert them to a logarithm (to base 10) of their value (pK_a), as with $[H^+(aq)]$ and pH.

Consider water at 298 K. The $[H^+(aq)]$ present is 10^{-7} mol dm^{-3}:

$$pH = \log\,[H^+(aq)]$$

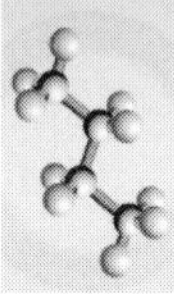

$pH = -\log(10^{-7}) = 7$

The numbers are not always this simple.

Suppose that in a given solution $[H^+(aq)] = 8.6 \times 10^{-9}\,mol\,dm^{-3}$.

To calculate the pH of this solution you need to understand how to take the logarithm of this sort of number:

$pH = -\log [H^+(aq)]$

$= -\log (8.6 \times 10^{-9})$

$= -((\log 8.6) - 9)$

$= -(0.9345 - 9)$

$= 8.0655$ or 8.07 to 3 s.f.

We can also calculate $[H^+(aq)]$ given the pH of a solution. Suppose there is an acidic solution with a pH of 2.73 and we want to calculate its hydrogen ion concentration:

$pH = 2.73 = -\log [H^-(aq)]$

$\log [H^-(aq)] = -2.73$

$= 1.862 \times 10^{-3}$

$= 1.86 \times 10^{-3}$ to 3 s.f.

Another piece of information that is helpful is the **ionic product** of water, K_W. This is the product of the $[H^+(aq)]$ and $[OH^-(aq)]$, and at 298 K is equal to $10^{-14}\,mol^2\,dm^{-6}$. This enables us to calculate $[OH^-(aq)]$ as well as $[H^+(aq)]$ in aqueous solutions.

Choosing indicators for titrations

On carrying out a titration it is the change in pH of the solution at the end point that is critical when choosing a suitable indicator. Two main factors need to be considered when choosing an indicator:

- The colour change should be sharp, i.e. no more than one drop of acid or alkali should give a distinct colour change.
- The end point should occur when the solution contains the same number of hydrogen ions as hydroxide ions.

Plotting graphs of the change in pH with the addition of the titrating solution, produces different curves for combinations of weak and strong acids with weak and strong bases, as Figure 7.2 shows.

Figure 7.2(a) shows the change in pH during the titration of a $25\,cm^3$ sample of $0.1\,mol\,dm^{-3}$ strong acid when adding strong base. Notice that there is a long vertical portion at $25\,cm^3$ showing a large change in pH.

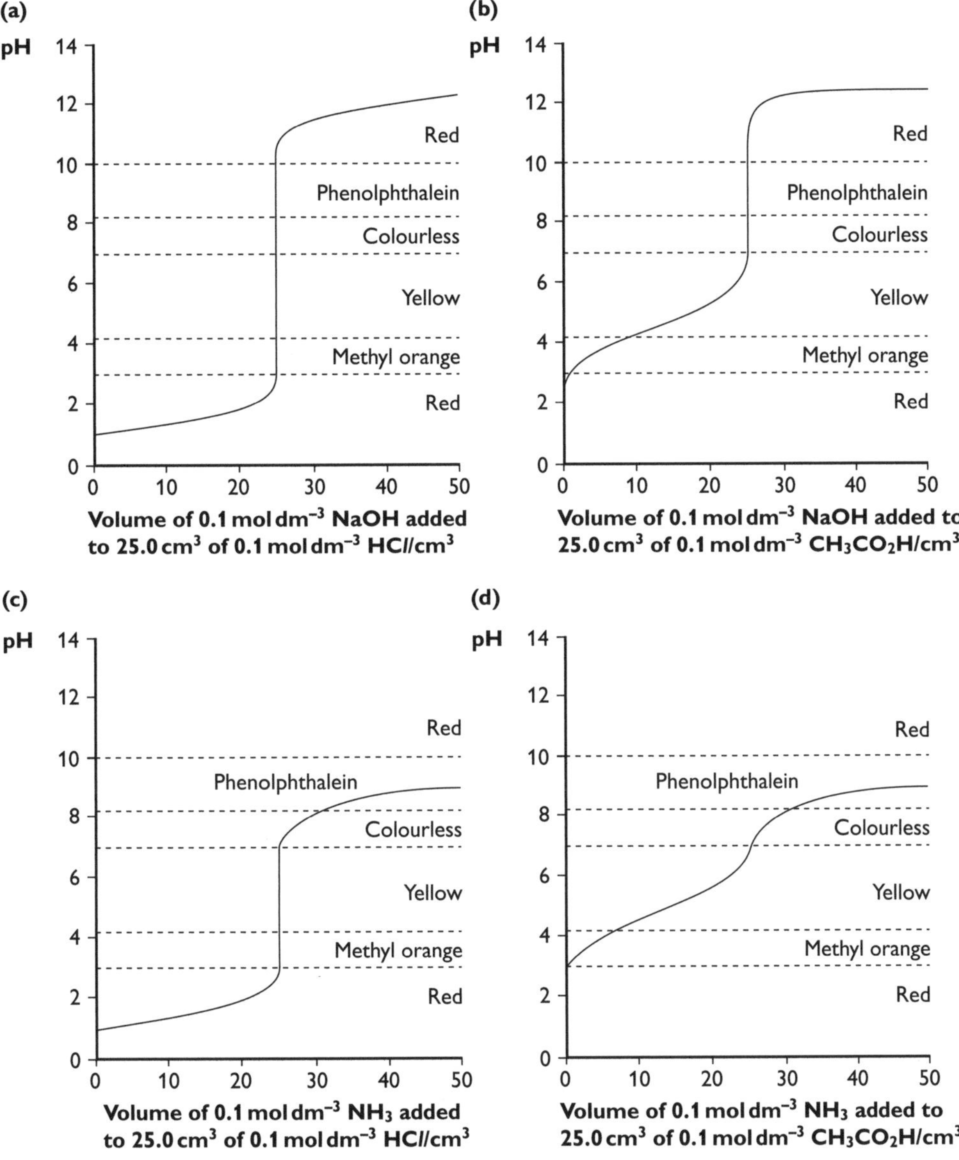

Figure 7.2(a) Titration of a strong acid with a strong base
(b) Titration of a weak acid with a strong base
(c) Titration of a strong acid with a weak base
(d) Titration of a weak acid with a weak base

Figure 7.2(b) shows the change in pH during the titration of a 25 cm^3 sample of 0.1 mol dm^{-3} weak acid when adding strong base. Notice that although the vertical portion occurs again at 25 cm^3, it is shorter and it starts at a higher pH.

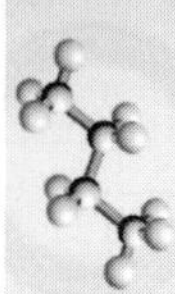

Figure 7.2(c) shows the change in pH during the titration of a 25 cm^3 sample of 0.1 mol dm^{-3} strong acid when adding weak base. Notice that although the vertical portion occurs again at 25 cm^3, it is shorter and it starts at a lower pH.

Figure 7.2(d) shows the change in pH during the titration of a 25 cm^3 sample of 0.1 mol dm^{-3} weak acid when adding weak base. Notice that now the vertical portion has practically disappeared, but it is closer to pH 7.

The diagrams also show the effective ranges of two common indicators — phenolphthalein and methyl orange. Table 7.2 shows the choice of indicator for the four titrations shown in Figure 7.2. Study the diagrams and try to decide on the reasons for the choices.

Tip The vertical portion has to be in the range to give a colour change.

Table 7.2

Acid/base combination	Indicator used
Strong acid/strong base	Either would do
Weak acid/strong base	Phenolphthalein
Strong acid/weak base	Methyl orange
Weak acid/weak base	Neither is suitable; a different method is needed

Buffer solutions

Buffer solutions are able to resist a change in acidity or alkalinity, maintaining an almost constant pH when a small amount of either substance is added. One important example of such a system occurs in blood, the pH of which is kept at around 7.4 by the presence of hydrogencarbonate ions, HCO_3^-.

Buffers rely on the dissociation of weak acids. Consider the weak acid HA, which dissociates as follows:

$$HA(aq) \rightleftharpoons H^+(aq) + A^-(aq)$$

Since it is a weak acid, $[H^+(aq)] = [A^-(aq)]$ and is very small. If a small amount of alkali is added, the OH^- ions react with the H^+ ions, removing them from the solution as water molecules. This disturbs the equilibrium. By Le Chatelier's principle, more HA will dissociate to restore the equilibrium and thus maintaining the pH. On the other hand, if we add acid, the H^+ ions will react with the A^- ions forming more HA. Unfortunately $[A^-(aq)]$ is very small and would soon be used up. This problem is solved by adding more A^- ions in the form of a salt of the acid HA, such as Na^+A^-. There would, therefore, be plenty of A^- ions to mop up any added H^+ ions. So a **buffer solution** consists of a weak acid and a salt of that weak acid.

In the case of the buffer system in blood, the equilibrium is:

$$H^+(aq) + HCO_3^-(aq) \rightleftharpoons CO_2(aq) + H_2O(l)$$

Addition of H^+ ions moves the equilibrium to the right, forming more carbon dioxide and water, while the addition of OH^- ions removes H^+ ions, causing the equilibrium to move to the left releasing more H^+ ions.

To calculate the pH of a buffer we can use the expression:

$$pH = pK_a + \log \frac{[salt]}{[acid]}$$

In other words, we need to know the pK_a of the acid, together with the concentration of the acid and its salt in the solution.

Example

Calculate the pH of an ethanoate–ethanoic acid buffer made by mixing 25 cm^3 of 0.100 mol dm^{-3} sodium ethanoate solution with 25 cm^3 of 0.100 mol dm^{-3} ethanoic acid solution. K_a for ethanoic acid is 1.8×10^{-5} mol dm^{-3}.

Answer

Mixing the two solutions means that the total volume is 50 cm^3, so the concentration of each is halved. Substituting these values in the equation gives:

$$pH = -\log (1.8 \times 10^{-5}) + \log \frac{(0.050)}{(0.050)}$$

$$= -\log (1.8 \times 10^{-5}) + \log 1$$

$$= -\log (1.8 \times 10^{-5}) \text{ (since } \log 1 = 0)$$

$$= 4.7$$

Solubility product

A final application of equilibria is in looking at the solubility of sparingly soluble salts. All the applications looked at so far have been in homogeneous equilibria (all the substances in the same phase). With solubility there are heterogeneous equilibria to consider, with one component in the solid phase and the remainder in the aqueous phase.

For example, in a saturated solution of silver chloride the following equilibrium exists:

$$AgCl(s) \rightleftharpoons Ag^+(aq) + Cl^-(aq)$$

The equilibrium constant for this system can be written as

$$K_c = \frac{[Ag^+(aq)][Cl^-(aq)]}{[AgCl(s)]}$$

However, it is not possible to change the concentration of the solid, so a new equilibrium constant is defined that allows for this. This is called the solubility product, K_{sp}, and is the product of the concentrations of the ions present in solution:

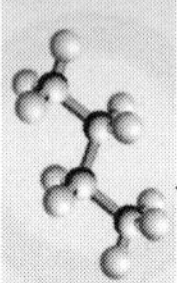

K_{sp} = [Ag^+(aq)][Cl^-(aq)]

In this case, the units of K_{sp} are $mol^2\,dm^{-6}$.

Example

Suppose we want to know if a precipitate will form when we mix equal quantities of a solution of silver nitrate and a solution of potassium chloride. We need to know the concentrations of the ions in each solution in $mol\,dm^{-3}$ and K_{sp} for silver chloride.

Let's assume that the concentration of potassium chloride is $1.0 \times 10^{-3}\,mol\,dm^{-3}$ and that of silver nitrate is $1.0 \times 10^{-5}\,mol\,dm^{-3}$. The K_{sp} of silver chloride at 298 K is $1.8 \times 10^{-10}\,mol^2\,dm^{-6}$.

Answer

On mixing equal quantities of the two solutions, each concentration is halved. Substituting the numbers into the expression for K_{sp} gives:

K_{sp} = [Ag^+(aq)][Cl^-(aq)]

[Ag^+(aq)] [Cl^-(aq)] = $5.0 \times 10^{-6} \times 5.0 \times 10^{-4}\,mol^2\,dm^{-6}$

= $2.5 \times 10^{-9}\,mol^2\,dm^{-6}$

This is greater than K_{sp} so a precipitate will form.

This same method can be used to calculate the concentration of one ion if we know that of the other ion and the K_{sp}. It is important to remember to write out the equilibrium because not all salts have a 1 : 1 ratio of ions.

If a substance is added that has an ion in common with a sparingly soluble salt, the concentration of that ion affects the equilibrium. This is known as the **common ion effect**. So, if sodium chloride solution is added to a saturated solution of silver chloride more solid is precipitated. This is because the extra chloride ions react with silver ions in solution. The silver chloride has become less soluble in the new mixture.

8 Reaction kinetics

This first part of this chapter deals with material needed for the AS exam, while the second part is only needed for the A2 exam.

Simple rate equations

You know from your practical work that the rate of a chemical reaction is affected by three main conditions:

- temperature
- concentration
- presence of a catalyst

At AS you need to be able to explain the effects of changes in these conditions using collision theory. It is important that you learn the correct terms to use when describing how reactions are influenced.

A reaction cannot take place unless the reacting particles collide with sufficient energy. Not all collisions result in reaction and the minimum energy required is called the **activation energy, E_a**. Increasing the temperature will increase the proportion of successful collisions. Increasing the concentration increases the chance of collisions taking place. In the presence of a catalyst, a reaction has a different mechanism, i.e. one of lower activation energy giving more successful collisions. At AS you need to be able to link these observations to the distribution of molecular energies and to explain the effects.

Boltzmann distribution of energies

The energy of molecules is directly proportional to their absolute temperature. The graph in Figure 8.1 shows a typical distribution of energies at constant temperature. This is known as the Boltzmann distribution.

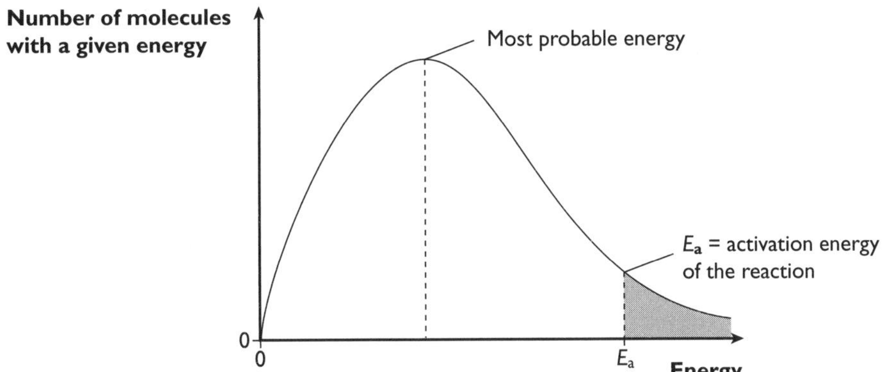

Figure 8.1 Boltzmann distribution

There are a number of points to remember about this graph:

- The distribution always goes through the origin.
- The curve approaches the x-axis but does not touch it.
- The peak represents the most probable energy.
- The area under the curve represents the total number of particles.
- E_a represents the activation energy (the minimum energy needed for reaction). The shaded portion represents the number of particles with energy greater than or equal to the activation energy ($E \geq E_a$).

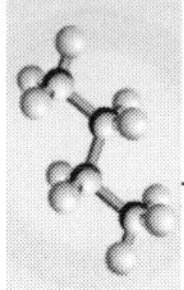

The effect of temperature

An increase in temperature changes the shape of the Boltzmann distribution curve as shown in Figure 8.2.

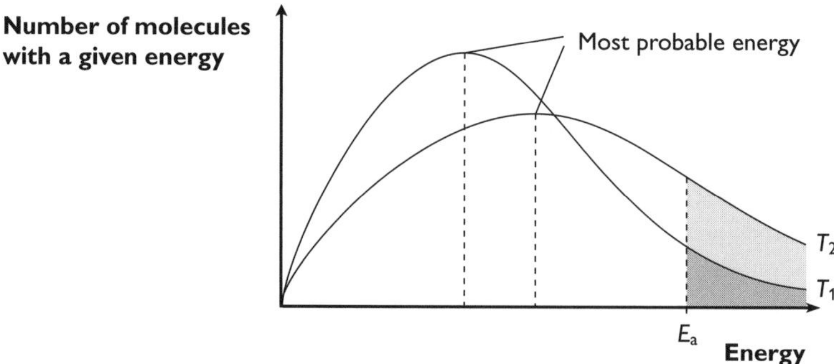

Figure 8.2 Boltzmann distribution at different temperatures

Notice that only the temperature has changed, so the area under the two curves is the same. The graph shows that at the higher temperature, T_2:

- there are fewer particles with lower energy (the curve is flatter)
- the most probable energy is higher
- more particles have $E \geq E_a$

So, at higher temperature a greater proportion of the particles have sufficient energy to react, and hence the rate of reaction increases. The reverse is true at a lower temperature.

The effect of concentration

The Boltzmann distribution is not relevant here. The explanation given at the beginning of the chapter in terms of increasing the chances of collision is adequate. It is worth remembering that increasing the pressure of a gas phase reaction has the same effect as increasing concentration in the liquid phase.

The effect of a catalyst

You should remember that catalysts speed up chemical reactions without being permanently changed themselves. In the presence of a catalyst a reaction has a different mechanism of activation energy, E_{cat}. This is shown in Figure 8.3.

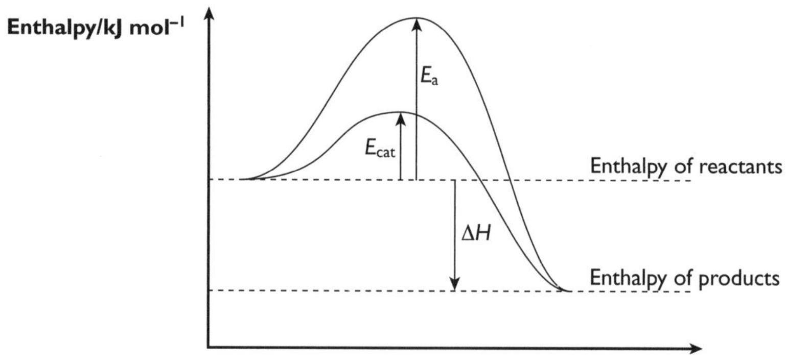

Figure 8.3 Activation energy in the presence and absence of a catalyst

It is important to remember that catalysts do not change the Boltzmann distribution for the temperature concerned. The position of E_a simply moves to the left increasing the proportion of particles with $E \geq E_a$ as shown in Figure 8.4.

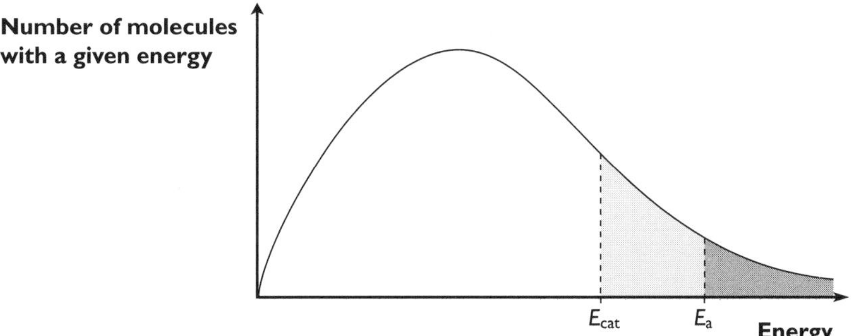

Figure 8.4 Position of E_a in the presence and absence of a catalyst

Enzymes are a particular type of catalyst found in biological systems. They are protein molecules and are specific in terms of the reaction they catalyse (see also page 178).

In the A2 examination you need to be able to manipulate data about rates of reaction in a more mathematical way. Some of this will come from practical work, or from data based on practical work.

Orders of reaction

When the rate relationships in a reaction are described mathematically, the rate equation is in the form:

$$\text{rate} = k[A]^m[B]^n$$

Here, k is the **rate constant** for the reaction between substances A and B, and m and n are the powers to which the concentrations of these substances are raised in the experimentally determined rate equation. They are also known as the **order** with respect to each substance. For the reactions you will come across at A2, m and n can be 0, 1 or 2.

It is important to remember that not all reactions take place in a single step. For multi-step reactions one step will always be slower than all the others. This is called the **rate-determining step**, since it is on this that the overall reaction rate depends.

It is easier to understand what this means by looking at some examples. Consider the reaction:

$$A + B \rightarrow \text{products}$$

If we measure the way in which the rate of this reaction changes with the concentrations of A and B, we might find that if we double the concentration of A then the rate doubles. We might also find that if we double the concentration of B then the rate also doubles.

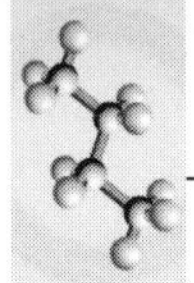

This tells us that the order with respect to A is 1, and that the order with respect to B is also 1; the overall order of the reaction is 2. The rate equation could be written as:

rate = $k[A]^1[B]^1$

However, we do not need to show m and n when they equal 1, so:

rate = $k[A][B]$

It is possible that in this reaction the rate does not depend on B reacting with A, but on A breaking down and B then reacting with the products. When this is the case, the rate equation is:

rate = $k[A][B]^0$

Remember that anything to the power 0 equals 1, so this rate equation should be written as:

rate = $k[A]$

Deducing order by the initial rates method

Most kinetic studies are based on experimental work. Consider a reaction for which the rate of reaction can be measured at the start. Data from such an experiment are given in Table 8.1.

Table 8.1

Run	Initial [A]/mol	Initial [B]/mol	Initial rate/mol s^{-1}
1	1.00	1.00	1.25×10^{-2}
2	1.00	2.00	2.5×10^{-2}
3	2.00	2.00	2.5×10^{-2}

Look at runs 1 and 2. If we double the concentration of B and keep the concentration of A constant, then the rate doubles. Look at runs 2 and 3. Doubling the concentration of A and keeping the concentration of B constant has no effect on the rate. This tells us that the reaction is first order with respect to B and zero order with respect to A. In other words, A does not feature in the rate equation. We can now calculate the rate constant:

rate = $k[B]$

2.5×10^{-2} mol s^{-1} = $k \times 2.0$ mol

Hence $k = 1.25 \times 10^{-2}\,s^{-1}$

Deducing order from graphs

Another way of deducing the order of a reaction with respect to a given reagent is to look at the graph of concentration against time. Zero-order (Figure 8.5(a)) and first-order reactions (Figure 8.5(b)) have characteristic shapes.

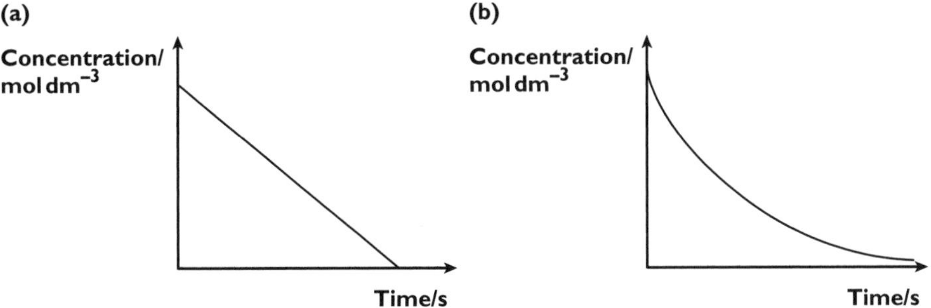

Fig 8.5 Concentration–time graph for (a) a zero-order reaction and (b) a first-order reaction

It is also possible to compare graphs of rate against concentration (Figure 8.6).

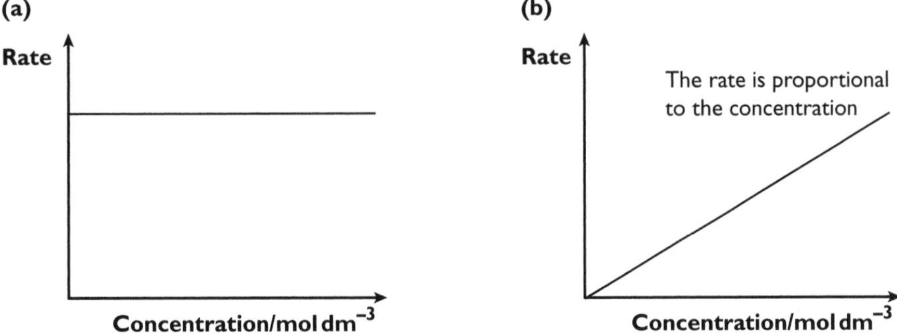

Figure 8.6 Rate–concentration graph of (a) a zero-order reaction and (b) a first-order reaction

The graphs show that in a zero-order reaction the rate is independent of concentration and gives a horizontal line (Figure 8.6(a)). For the first-order reaction, the reaction shows a constant time for the concentration of reactant to halve (Figure 8.5(b)). This is known as the **half-life** of the reaction and is similar to the half-life used in radioactive decay.

The half-life can be used to calculate the rate constant of a reaction. Look at the following reaction:

$$CH_3N_2CH_3(g) \rightarrow CH_3CH_3(g) + N_2(g)$$

When the compound is heated it decomposes into the two gases shown. No other reactants are needed so:

$$\text{rate} = k[CH_3N_2CH_3(g)]$$

The half-life at 500 K is about 1750 s. This means that if the starting concentration is 0.10 mol dm^{-3}, after 1750 s the concentration will have halved to 0.05 mol dm^{-3}. In another 1750 s, the concentration will have halved to 0.025 mol dm^{-3}, and so on.

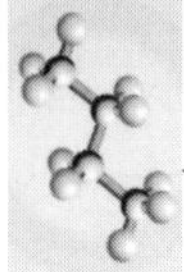

Half-life can be used to calculate the rate constant, k. For a first-order reaction:

$$k = \frac{0.693}{t_{1/2}} = \frac{0.693}{1750} = 3.96 \times 10^{-4}\,s^{-1}$$

For the examination you need to be able to predict the order of a reaction from a given mechanism (and vice versa).

Consider the following reaction:

$C_4H_9Br + OH^- \rightarrow C_4H_9OH + Br^-$

This could occur in one step as shown above or in two steps:

slow step: $C_4H_9Br \rightarrow C_4H_9^+ + Br^-$

fast step: $C_4H_9^+ + OH^- \rightarrow C_4H_9OH$

If the first mechanism is correct we would predict that the reaction is first order with respect to C_4H_9Br and to OH^-. If the second mechanism is correct we would predict that the reaction is first order only with respect to C_4H_9Br, since this is the rate-determining step. Practical evidence suggests that the first mechanism is correct.

Experimental techniques for studying rates

It is important to consider the methods you could use to follow the progress of a reaction if you want to determine the rate.

Sampling followed by titration

Small amounts of the reaction mixture are withdrawn by pipette at regular intervals and the concentration of one of the reactants or products is determined by titration. Two examples are the formation of an acid and an iodination reaction.

Using a colorimeter

This method only works if one of the reactants or products is coloured. It has advantages over titration in that no sampling is needed and it gives an almost instantaneous result. An example is the formation of a transition metal complex.

Measurement of gas evolved

One of the products has to be a gas for this method to work. The volume of gas is measured in a syringe or by the displacement of water from an upturned burette. An example is the reaction between an acid and a carbonate.

Catalysis

You will already have come across the use of catalysts at GCSE and at AS. At A2 you need to know how catalysts are able to speed up reactions. Four specific examples are mentioned in the syllabus.

The Haber process

You saw in the previous chapter how this equilibrium process was made economic by the use of an iron catalyst. The catalyst is heterogeneous — in a different phase from the gases. Transition metals are particularly good at acting as catalysts because their atoms have unfilled d-orbitals. The gases are adsorbed on to the surface of the metal, forming weak bonds (Figure 8.7). This can have one of two consequences:

- The formation of bonds with the metal surface may weaken the bonds within the gas molecules.
- The orientation of the adsorbed molecules may be favourable for the reaction.

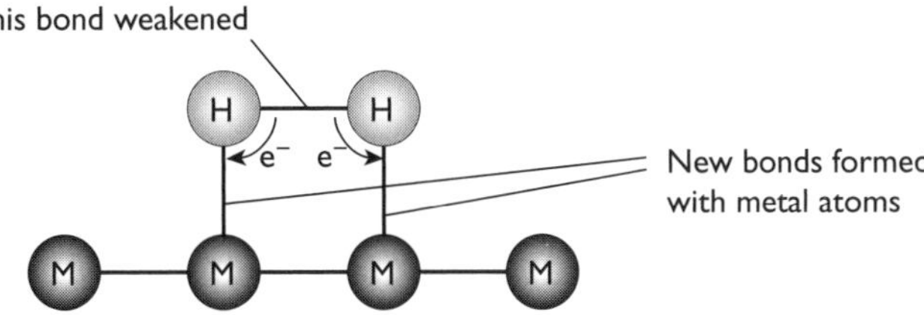

Figure 8.7 Adsorption of a gas onto the surface of a metal catalyst

Catalytic converters in vehicle exhausts

Catalytic converters have become important in recent years and are designed to remove a number of pollutant gases from vehicle exhausts. The problem is complex because some pollutants, such as carbon monoxide, have to be oxidised while others, such as nitrogen oxides, have to be reduced.

The converter consists of a ceramic honeycomb with a very thin coat of platinum, rhodium and palladium (all expensive metals). The platinum and rhodium help reduce the NO_x to nitrogen while platinum and palladium help oxidise the CO and unburnt hydrocarbons. The car has to run on unleaded petrol because lead would 'poison' the catalyst, making it ineffective. This again is a heterogeneous system.

Nitrogen oxides in the atmosphere

This is an example of homogeneous catalysis — the reactants and the catalyst are in the same phase, in this case gases. Studies on acid rain have concluded that in the atmosphere the presence of oxides of nitrogen, particularly nitrogen(IV) oxide, NO_2, increases the rate of oxidation of sulfur dioxide to sulfur trioxide. Nitrogen(IV) oxide is unchanged by the reaction and is thought to form a weak intermediate with the sulfur dioxide.

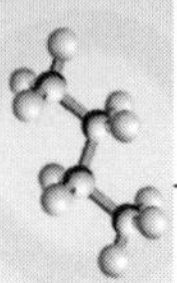

The role of Fe^{2+} in the $I^-/S_2O_8^{2-}$ reaction

The oxidation of iodide ions by peroxodisulfate ions is another example of homogeneous catalysis. In this case, all the species are in the aqueous phase. It is believed that in the presence of Fe^{3+} ions this oxidation occurs in two steps:

Overall reaction: $S_2O_8^{2-}(aq) + 2I^-(aq) \rightarrow 2SO_4^{2-}(aq) + I_2(aq)$

Catalysed reaction: $S_2O_8^{2-}(aq) + 2Fe^{2+}(aq) \rightarrow 2SO_4^{2-}(aq) + 2Fe^{3+}(aq)$

$2Fe^{3+}(aq) + 2I^-(aq) \rightarrow 2Fe^{2+}(aq) + I_2(aq)$

Although there are two steps in the reaction, the overall activation energy is lower than in the single-step reaction, as can be seen in Figure 8.8.

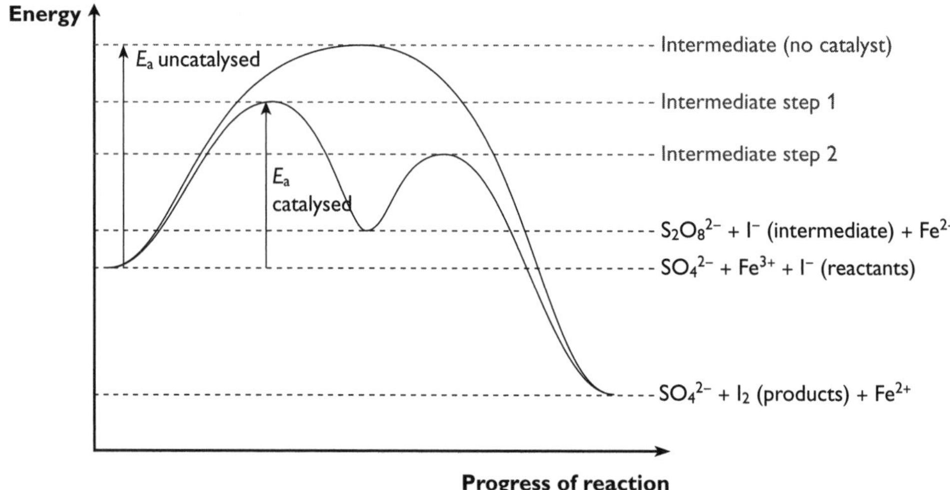

Figure 8.8 Effect of a catalyst on the reaction profile

9 Chemical periodicity

All of the material in this chapter is needed for the AS exam.

Physical properties of elements

You need to know and be able to explain the variation of four key physical properties of elements across the third period (sodium to argon). These are:

- atomic and ionic radius
- melting point
- electrical conductivity
- ionisation energy

Atomic and ionic radius

The periodic table is an arrangement of the chemical elements according to their proton number. The word 'periodic' suggests a regular recurrence of a feature. Look at the graph of atomic volume (which is linked directly to atomic radius) shown in Figure 9.1. A pattern is apparent — the Group I metals occur at peaks on the graph.

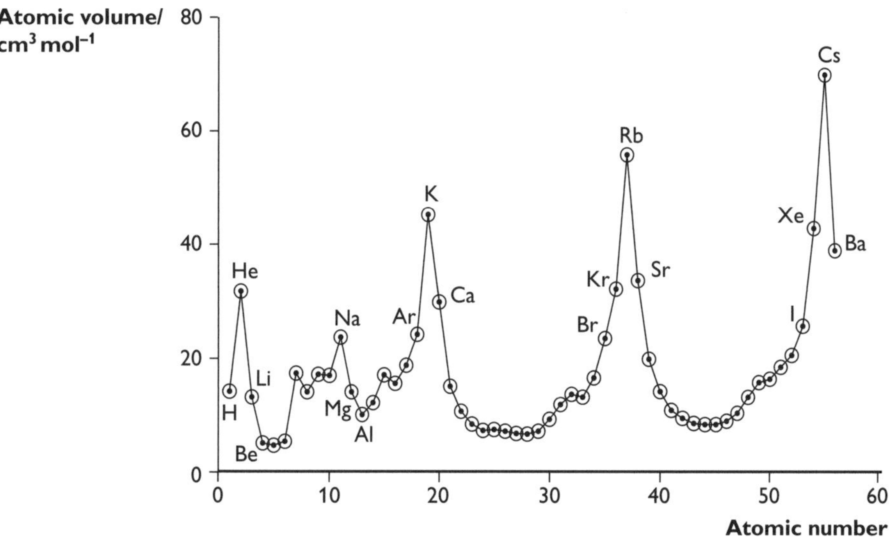

Figure 9.1 Relationship between atomic volume and atomic number

The other thing that is noticeable is that atoms of the Group I metals get larger moving down the group.

Moving across period 3 from left to right, the size of the ionic radii changes (Figure 9.2). To begin with, elements lose electrons to form positive ions. The increasing positive charge pulls the electrons closer to the nucleus reducing the ionic radius. Beyond silicon, Si^{4+}, the atoms form ions with a decreasing negative charge. The electrons gained fill the next electron shell. This means that the ionic radius is much larger at phosphorus, P^{3-}, and then decreases slightly until chlorine, Cl^{-}.

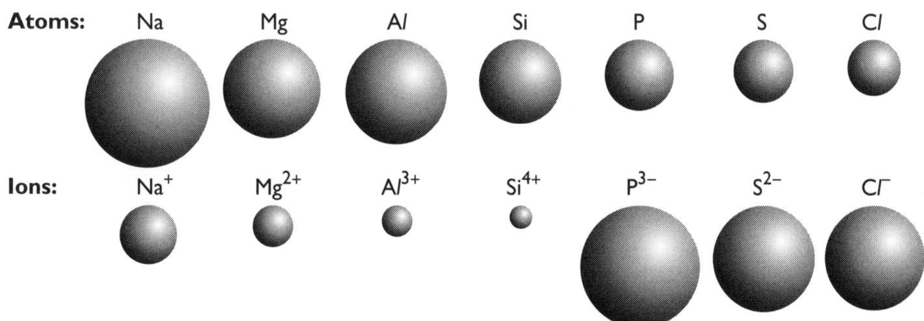

Figure 9.2 Atomic and ionic radii of period 3 elements

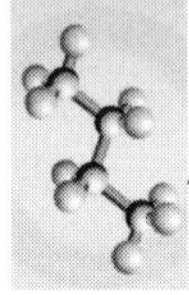

Melting point

Period 3 contains different types of elements, from metals on the left-hand side, through non-metallic solids to gases on the right-hand side. The melting and boiling points of these elements are shown in Figure 9.3.

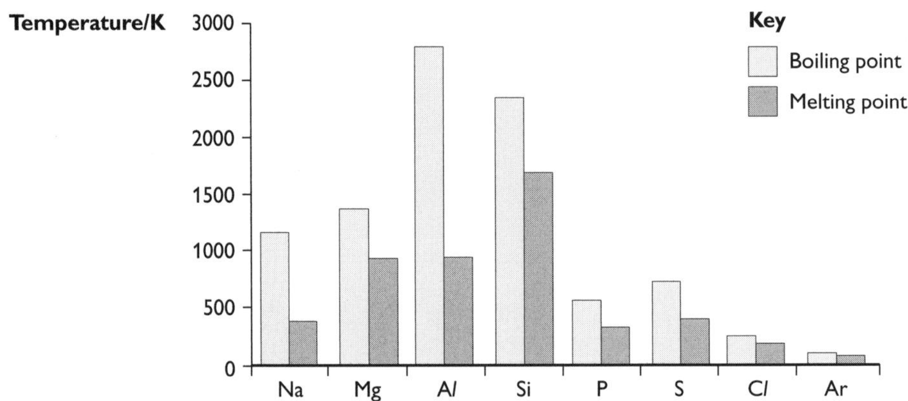

Figure 9.3 Melting and boiling points of period 3 elements

The elements sodium, magnesium and aluminium are metals and their atoms are bonded with a 'sea' of electrons. The melting (and boiling) points increase because the number of electrons each atom contributes increases. Silicon is a semi-conductor with a giant covalent structure (similar to that of diamond), so it has a high melting point. Phosphorus is a non-metal with four atoms in its molecules. To melt it, only the van der Waals forces have to be broken, so phosphorus has a low melting point. Sulfur, another non-metal, has eight atoms in its molecules. Since the molecules are bigger, there are larger van der Waals forces than in phosphorus. Hence, sulfur has a higher melting point than phosphorus. A chlorine molecule has only two atoms, so the melting point is lower than that of sulfur. Finally, argon consists of single atoms with very small van der Waals forces. Hence, argon has the lowest melting (and boiling) point in period 3.

Electrical conductivity

Sodium, magnesium and aluminium are good electrical conductors because of the 'sea' of delocalised electrons they possess. Silicon is a semi-conductor, but not as good a conductor as graphite. All the other elements are electrical insulators.

Ionisation energy

We discussed ionisation energy in Chapter 2. Figure 9.4 shows the change in first ionisation energy across period 3.

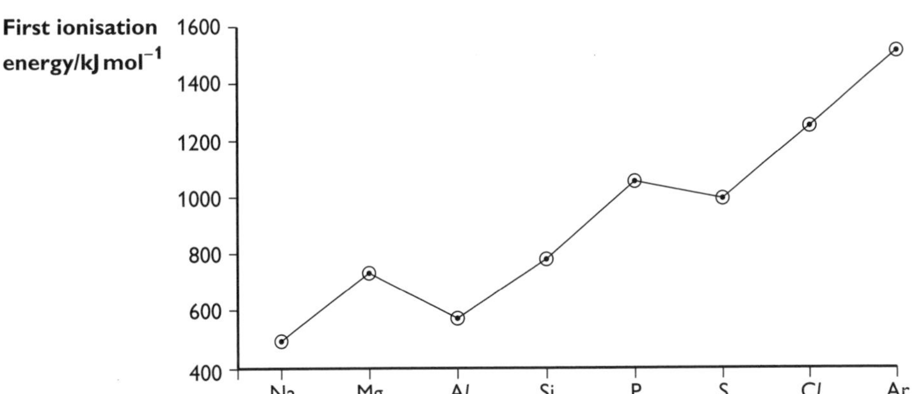

Figure 9.4 Change in first ionisation energy across period 3

There are four things that affect the size of the first ionisation energy:

- the charge on the nucleus
- the distance of the electron from the nucleus
- the number of electrons between the outer electrons and the nucleus
- whether the electron is alone or paired in its orbital

For period 3, the charge on the nucleus increases by one proton for each element. In each case the electrons are being removed from the third shell, and these are screened by the $1s^2$, $2s^2$ and $2p^6$ electrons. The graph is not a straight line because of the orbitals the electrons are removed from.

The first ionisation energy of aluminium is lower than that of magnesium. This drop is because in aluminium the electron removed is in a p-orbital and is, on average, further away from the nucleus than an electron in an s-orbital. There is another drop between phosphorus and sulfur. This drop is caused by removing a paired electron in a p-orbital. The repulsion between these two electrons makes it easier to remove one of them than a single electron in a p-orbital.

Chemical properties of elements

You have to be able to recall the reactions of these elements with oxygen, chlorine and water, and to know about the reactions of any oxides and chlorides formed with water.

Reactions with oxygen

You have probably seen most of the period 3 elements react with air, or perhaps with oxygen. The reactions are summarised in Table 9.1.

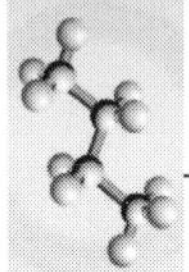

Table 9.1

Element	Reaction	Product(s)	Equation
Sodium	Burns with an orange-yellow flame to give white products	Sodium oxide and peroxide	$4Na + O_2 \rightarrow 2Na_2O$ $2Na + O_2 \rightarrow Na_2O_2$
Magnesium	Burns with a bright white flame to give a white product	Magnesium oxide	$2Mg + O_2 \rightarrow 2MgO$
Aluminium	Powder burns to give a white product	Aluminium oxide	$4Al + 3O_2 \rightarrow 2Al_2O_3$
Silicon	Burns if heated strongly	Silicon dioxide	$Si + O_2 \rightarrow SiO_2$
Phosphorus	Burns with a yellow flame producing clouds of white smoke	Phosphorus(III) oxide; phosphorus(V) oxide in excess O_2	$P_4 + 3O_2 \rightarrow P_4O_6$ $P_4 + 5O_2 \rightarrow P_4O_{10}$
Sulfur	Burns with a blue flame producing a colourless gas	Sulfur dioxide (sulfur trioxide is produced in the presence of a catalyst and excess O_2)	$S + O_2 \rightarrow SO_2$
Chlorine	Does not react directly with oxygen		
Argon	No reaction		

Reactions with chlorine

You may not have seen as many of the elements reacting with chlorine. The reactions are summarised in Table 9.2.

Table 9.2

Element	Reaction	Product	Equation
Sodium	Burns with a bright orange flame giving a white product	Sodium chloride	$2Na + Cl_2 \rightarrow 2NaCl$
Magnesium	Burns with a bright white flame giving a white product	Magnesium chloride	$Mg + Cl_2 \rightarrow MgCl_2$
Aluminium	Burns with a yellow flame giving a pale yellow product	Aluminium chloride	$2Al + 3Cl_2 \rightarrow 2AlCl_3$
Silicon	Reacts when chlorine gas is passed over it to form a colourless liquid	Silicon tetrachloride	$Si + 2Cl_2 \rightarrow SiCl_4$
Phosphorus	Burns with a yellow flame to form a mixture of chlorides	Phosphorus(III) chloride and phosphorus(V) chloride	$P_4 + 6Cl_2 \rightarrow 4PCl_3$ $P_4 + 10Cl_2 \rightarrow 4PCl_5$
Sulfur	Reacts when chlorine gas is passed over it to form an orange liquid	Disulfur dichloride	$2S + Cl_2 \rightarrow S_2Cl_2$
Chlorine	No reaction		
Argon	No reaction		

Reactions with water

Sodium reacts violently with water, releasing hydrogen gas and dissolving to form sodium hydroxide solution:

$$2Na + 2H_2O \rightarrow 2NaOH + H_2$$

Magnesium reacts slowly with cold water, forming magnesium hydroxide and bubbles of hydrogen:

$$Mg + 2H_2O \rightarrow Mg(OH)_2 + H_2$$

It reacts vigorously if steam is passed over the heated metal, forming magnesium oxide and hydrogen:

$$Mg + H_2O \rightarrow MgO + H_2$$

Oxidation numbers in oxides and chlorides

Figure 9.5 shows a plot of the oxidation numbers in the oxides and chlorides of period 3 elements.

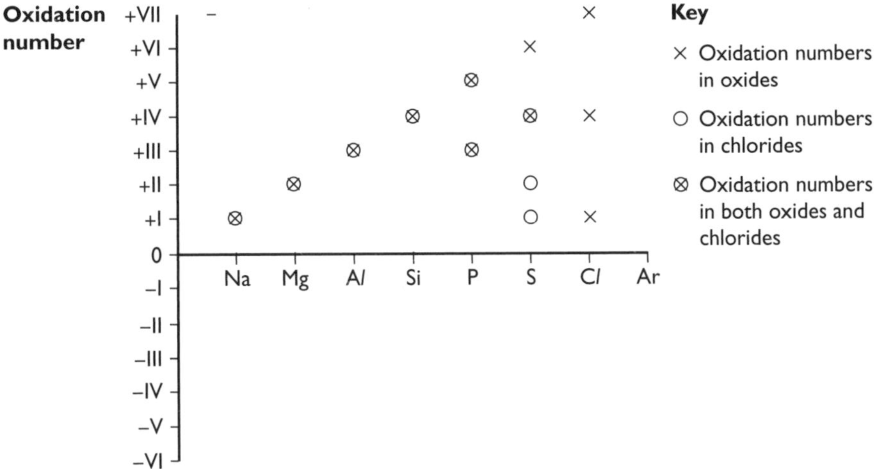

Figure 9.5 Oxidation numbers of the oxides and chlorides of period 3 elements

The first four elements in the period show positive oxidation numbers that correspond to the loss of all their outer electrons (silicon can also gain four electrons in forming its hydride, SiH_4). Elements in groups 5, 6 and 7 can also show positive oxidation numbers in their oxides and chlorides. You may think that it is unusual for non-metals to have positive oxidation numbers, but carbon has a positive oxidation number in carbon dioxide.

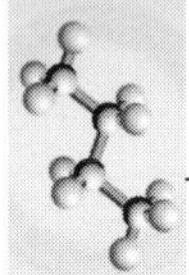

Reactions of oxides with water and their acid–base behaviour

The general trend to remember is that alkalis are formed on the left-hand side of a period, aluminium and silicon oxides are almost insoluble, and acids are formed on the right-hand side (Table 9.3).

Table 9.3

Oxide	Reaction	pH of solution	Equation
Sodium	Dissolves exothermically	14	$Na_2O + H_2O \rightarrow 2NaOH$
Magnesium	Slight reaction	9	$MgO + 2H_2O \rightarrow Mg(OH)_2$
Aluminium	No reaction		
Silicon	No reaction		
Phosphorus	Phosphorus(III) oxide reacts with cold water Phosphorus(V) oxide reacts violently with water	1–2 1–2	$P_4O_6 + 6H_2O \rightarrow 4H_3PO_3$ $P_4O_{10} + 6H_2O \rightarrow 4H_3PO_4$
Sulfur	Sulfur dioxide dissolves readily in water Sulfur trioxide reacts violently with water	1 0	$SO_2 + H_2O \rightarrow H_2SO_3$ $SO_3 + H_2O \rightarrow H_2SO_4$
Chlorine	Does not react directly with oxygen		
Argon	No reaction		

Aluminium oxide is **amphoteric**. This means that it reacts with both acids and alkalis. Aluminium oxide contains oxide ions, so it reacts as a base with acids in a similar way to magnesium oxide:

$$Al_2O_3 + 6HCl \rightarrow 2AlCl_3 + 3H_2O$$

It also has a significant acidic characteristic, reacting with alkalis, such as sodium hydroxide, to form an aluminate:

$$Al_2O_3 + 2NaOH + 3H_2O \rightarrow 2NaAl(OH)_4$$

Reactions of the chlorides with water

The reactions of the chlorides of period 3 elements with water (Table 9.4) give a clue to the bonding present. This is linked to the electronegativity of the element. As you saw in Chapter 3, the larger the difference in electronegativity, the more polar is the bond. So sodium and chlorine have a difference of 2.1 whereas sulfur and chlorine have a difference of only 0.5.

Table 9.4

Chloride	Bonding	Electro-negativity	Reaction	Equation
Sodium	Ionic (electrovalent)	0.9	Dissolves to give Na^+ and Cl^- ions	
Magnesium	Ionic (electrovalent)	1.2	Dissolves to give Mg^{2+} and Cl^- ions	
Aluminium	Mainly covalent	1.5	Hydrolyses	$AlCl_3 + 3H_2O \rightarrow Al(OH)_3 + 3HCl$
Silicon	Covalent	1.8	Hydrolyses	$SiCl_4 + 2H_2O \rightarrow SiO_2 + 4HCl$
Phosphorus	Covalent	2.1	Hydrolyses	$PCl_3 + 3H_2O \rightarrow H_3PO_3 + 3HCl$
Sulfur	Covalent	2.5	Hydrolyses	$2S_2Cl_2 + 2H_2O \rightarrow SO_2 + 4HCl + 3S$
Chlorine	No reaction	3.0		
Argon	No reaction			

Try this yourself

(26) An element J in period 3 reacts vigorously with oxygen when in powder form. It forms a liquid chloride that hydrolyses in water to give an acidic solution and an insoluble solid.

Identify J and write balanced equations for the three reactions described.

10 Group chemistry

You need to know about the chemistry of three groups in the periodic table —Groups II, IV and VII — as well as the transition metals, which are covered in Chapter 11, and the elements nitrogen and sulfur, which are covered in Chapter 12.

Groups II and VII are needed mainly for AS, and group IV for A2.

Group II

Physical properties of the elements

Some of the most useful physical properties of the Group II metals are shown in Table 10.1.

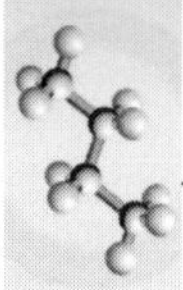

Table 10.1

Element	Atomic radius/nm	1st ionisation energy/kJ mol^{-1}	Electronegativity	Melting point/°C
Beryllium	0.111	900	1.57	1278
Magnesium	0.160	738	1.31	649
Calcium	0.197	590	1.00	839
Strontium	0.215	550	0.95	769
Barium	0.217	503	0.89	729

The first three physical properties show steady trends — upwards in atomic radius and downwards in first ionisation energy and electronegativity. The decrease in melting point would fit this pattern if it were not for the anomalous low value for magnesium.

We can explain the change in atomic radius in terms of the additional shell of electrons added for each period and the reduced effective nuclear charge as more electron shells are added.

You need to remember that ionisation energy is governed by:

- the charge on the nucleus
- the amount of screening by the inner electrons
- the distance between the outer electrons and the nucleus

As Group II is descended, the increase in charge on the nucleus is offset by the number of inner electrons. However, the distance of the outer electrons from the nucleus increases and the first ionisation energy decreases down the group.

Electronegativity is the ability of an atom to attract a pair of bonding electrons. As the size of the atom increases, any bonding pair of electrons is further from the nucleus, which means it is less strongly held and the electronegativity falls. The effect of this is to increase the ionic (electrovalent) character of any compounds formed as the group is descended.

Reactions of the elements with oxygen and water

The only reactions you need to learn for the Group II elements are their reactions with oxygen and water. These are summarised in Table 10.2.

Table 10.2

Element	Reaction with O_2	Reaction with H_2O
Beryllium	Reluctant to burn, white flame	No reaction
Magnesium	Burns easily with a bright white flame	Reacts vigorously with steam but only very slowly in water
Calcium	Difficult to ignite, flame tinged with red	Reacts moderately with water forming the hydroxide
Strontium	Difficult to ignite, flame tinged with red	Reacts rapidly with water forming the hydroxide
Barium	Difficult to ignite, flame tinged with green	Reacts vigorously with water forming the hydroxide

The general equation for the reaction with oxygen is:

$$2X(s) + O_2(g) \rightarrow 2XO(g)$$

where X is any metal in the group.

Both strontium and barium can also form a peroxide as well as the oxide:

$$Y(s) + O_2(g) \rightarrow YO_2(g)$$

where Y is strontium or barium.

The general equation for the reaction with water is:

$$X(s) + 2H_2O(l) \rightarrow X(OH)_2(s) + H_2(g)$$

The exception to this is magnesium, which forms the oxide when reacted with steam.

Behaviour of the oxides with water

Beryllium oxide is amphoteric, but all the other oxides are sparingly soluble in water producing solutions of increasing base strength:

$$XO(s) + H_2O(l) \rightarrow X(OH)_2(aq)$$

Thermal decomposition of nitrates and carbonates

The general rule to remember is that the lower down the group a metal is, the more stable is its nitrate and carbonate.

This is a result of the ability of the cation to polarise the anion. This is greater at the top of the group, where the cations are smaller and possess a high charge density. This applies to both the nitrate and carbonate where polarisation results in the formation of the oxide:

$$2X(NO_3)_2(s) \rightarrow 2XO(s) + 4NO_2(g) + O_2(g)$$

$$XCO_3(s) \rightarrow XO(s) + CO_2(g)$$

We can examine this trend by comparing the decomposition temperatures of the carbonates (Table 10.3).

Table 10.3

Element	Decomposition temperature of the carbonate/K
Beryllium	Unstable at 298
Magnesium	700
Calcium	1200
Strontium	1580
Barium	1660

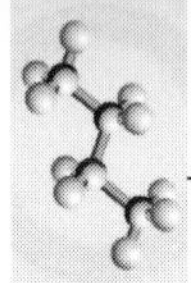

Important uses of Group II compounds

Magnesium oxide

Magnesium oxide has high stability coupled with a melting point of over 3000 K. This makes it particularly suitable for lining furnaces and other high temperature applications.

Calcium carbonate

Limestone is a widespread naturally occurring form of calcium carbonate. It has been used for thousands of years in the construction of buildings and roads. As well as a building material, it is used in the manufacture of cement. Marble is another form of calcium carbonate and is also used in building.

'Lime' is used to reduce the acidity of soils in agriculture. It consists of powdered calcium carbonate which neutralises the soil to provide better growing conditions for crops. The application of lime must be controlled carefully to achieve the right soil pH and not make it too alkaline.

Solubility of Group II sulfates

The solubility of the sulfates of Group II decreases down the group. This is due to a combination of the relative sizes of the enthalpy change of hydration of the cations and the lattice energy of the sulfate concerned.

As the cations get bigger, the energy released as the ions bond to water molecules (the enthalpy change of hydration) falls. Larger ions are not as strongly attracted to the water molecules. As you go down a group, the energy needed to break up the lattice decreases as the positive ions get bigger. The bigger the ions, the more distance there is between them and the weaker are the forces holding them together. Since both energy changes fall, it is a question of which is the more significant. For large ions such as sulfate, it is the enthalpy change of hydration factor that dominates.

Group IV

This group only applies to the A2 examination. Within Group IV there is a change from a non-metal at the top of the group (carbon), through the semiconductors (silicon and germanium) to metals (tin and lead) at the bottom of the group.

Melting point and electrical conductivity

Table 10.4

Element	Melting point/K	Electrical conductivity	Electronegativity
Carbon (diamond)	3930	Insulator	2.55
Silicon	1680	Semiconductor	1.90
Germanium	1210	Semiconductor	2.01
Tin	505	Good	1.96
Lead	601	Good	2.33

The melting points and electrical conductivities bear out what is known about the trend from non-metal to metal in this group. However, trying to *explain* this trend is much more difficult. If you look at the electronegativities in the Table 10.4, there is nothing to suggest a relationship.

In diamond, all the electrons are bound strongly in a three-dimensional lattice in which each carbon atom is bonded to four adjacent atoms (Figure 10.1).

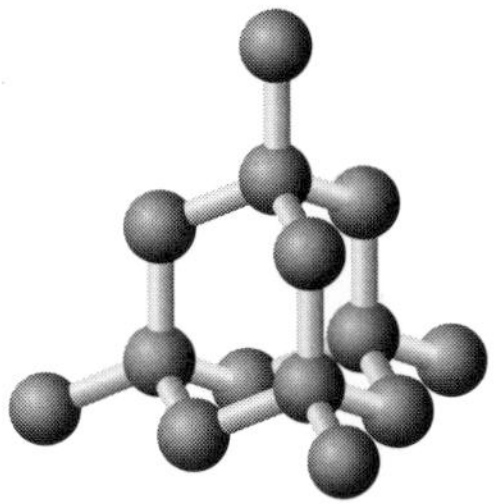

Figure 10.1 Structure of diamond

We also know that another form of carbon — graphite — is a good conductor of electricity, despite being a non-metal. In the case of graphite, this can be explained in terms of the structure, which consists of layers of carbon atoms joined in planes, with 'electron-rich' gaps between them that carry the current (Figure 10.2).

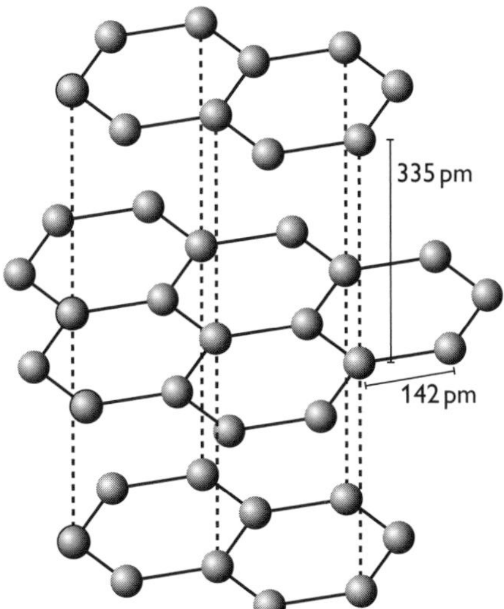

Figure 10.2 Structure of graphite

There is a change from purely covalent bonding in carbon to purely metallic bonding in lead. However, the reasons for this are far from clear-cut and you will not be expected to state these.

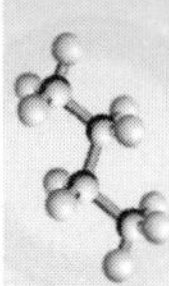

The tetrachlorides

All the elements form covalent tetrahedral chlorides of the form MCl_4, which are liquid at room temperature. However, the stability of the tetrachlorides varies considerably as we descend the group. For now, we will ignore the organic chlorides that carbon can form (see Chapter 15).

With the exception of carbon tetrachloride, CCl_4, which is stable in air and water, the tetrachlorides hydrolyse easily by accepting a lone pair of electrons from water as a first step.

Carbon has no available orbital to accept a lone pair and the chlorine atoms are so bulky that it would be almost impossible for the oxygen atom to get close to the carbon atom. With larger atoms further down the group, both of these difficulties can be overcome, for example:

$$SiCl_4(l) + 2H_2O(l) \rightarrow SiO_2(s) + 4HCl\ (g)$$

The situation with lead(IV) chloride is more complicated. This is because it decomposes at room temperature forming lead(II) chloride (which shows ionic (electrovalent) characteristics) and chlorine:

$$PbCl_4(l) \rightarrow PbCl_2(s) + Cl_2(g)$$

Oxides

For the exam you need to be able to describe and explain the bonding, acid–base characteristics and thermal stability of the oxides of Group IV elements in oxidation states 2 and 4. In practice, this is less complicated than it seems since silicon and germanium only show stable +4 oxidation states, and tin and lead behave in a similar way to each other.

Carbon forms two oxides, carbon monoxide and carbon dioxide. Carbon is different from other members of the group since it is able to form multiple bonds with oxygen.

Carbon dioxide is a linear molecule with two double covalent bonds formed to oxygen (Figure 10.3(a)). In carbon monoxide there is a double and a coordinate (or dative) bond (Figure 10.3(b)).

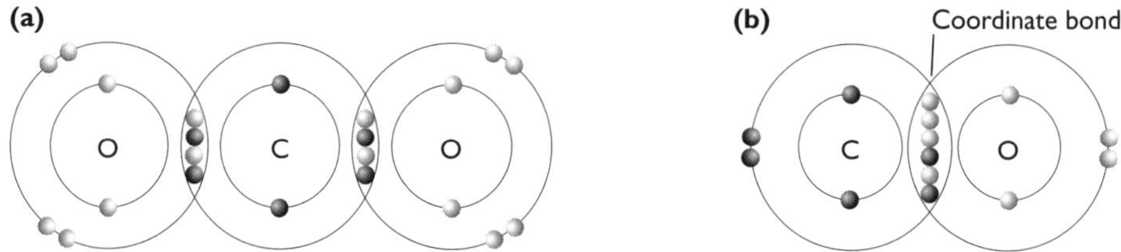

Figure 10.3 Bonding in (a) carbon dioxide and (b) carbon monoxide

By contrast, silicon dioxide consists of a giant lattice held together by silicon–oxygen single bonds (Figure 10.4). It is similar in overall shape to the structure of diamond.

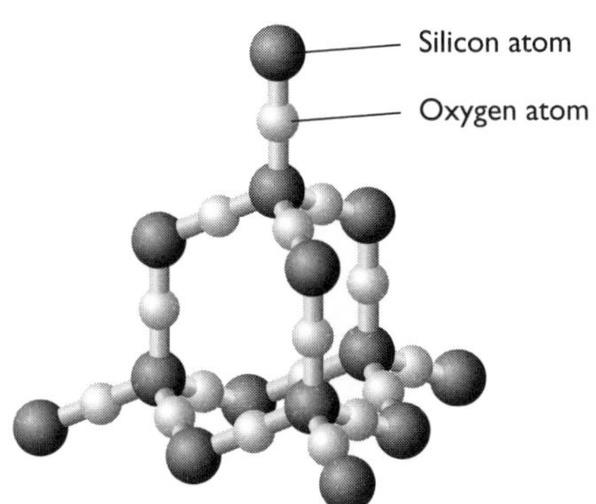

Figure 10.4 Structure of silicon(IV) oxide

Germanium(IV), tin(IV) and lead(IV) oxides all have giant structures. Tin(IV) and lead(IV) oxides are ionic (electrovalent). Tin and lead also form ionic (electrovalent) oxides in the +2 oxidation state.

As we descend the group there is a change in the acid–base character of the oxides. Carbon and silicon dioxides are acidic, germanium and tin dioxides are amphoteric and there is some evidence that lead dioxide is almost basic.

$$MO_2(s) + 4HCl(aq) \rightarrow MCl_4(l) + 2H_2O(l)$$

$$MO_2(s) + 2OH^-(aq) + 2H_2O(l) \rightarrow [M(OH)_6]^{2-}(aq)$$

where M is germanium or tin.

Carbon monoxide is often considered neutral, but it can react with hot concentrated sodium hydroxide solution. This suggests a degree of acidity. Tin and lead monoxides are amphoteric:

$$MO(s) + 2OH^-(aq) \rightarrow MO_2^{2-}(aq) + H_2O(l)$$

$$MO(s) + 2H^+(aq) \rightarrow M^{2+}(aq) + H_2O(l)$$

Of the dioxides, only lead dioxide is decomposed by heating at a moderate temperature:

$$2PbO_2(s) \rightarrow 2PbO(s) + O_2(g)$$

With the monoxides, only PbO is stable on heating in oxygen; the others are oxidised to the dioxide.

Relative stability of the oxidation states

We have already seen some instances of the relative stability of the +2 and +4 oxidation states in Group IV. It is important not only to remember that these differences exist, but to be able to explain them. In an exam question you might be given $E^\ominus$ data and asked for an interpretation in terms of the +2 and +4 states. Remember that the $E^\ominus$ data only apply to dilute aqueous solutions and not to solids.

In some textbooks you will read about the stability of the +2 state in tin and lead as being caused by the 'inert pair effect'. Just quoting this in an examination will not

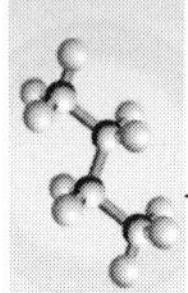

gain any marks because it does not explain anything. The term refers to the fact that for tin and lead the energy gap between the s- and p-electrons is sufficient to prevent easy promotion of an s-electron followed by hybridisation. In carbon, the energy required to promote an electron from the 2s to the 2p orbital is compensated by the energy released when four covalent bonds are formed. As we descend Group IV, the covalent bonds become longer and weaker as the atoms get larger. The energy released in forming the bonds is no longer big enough to compensate for the promotion of the 2s electron, and hence the +2 state becomes more favourable.

Uses of silicon dioxide

Silicon dioxide (in its impure form known as 'sand') is an extremely common compound on Earth. Over thousands of years it has been used in a number of ways, most commonly in ceramics. It is an important component of glass where it is mixed with sodium carbonate and calcium oxide together with magnesium and aluminium oxides. Glasses with special properties have other materials added.

Pottery has been produced for thousands of years by heating clay (which contains mixed silicates) to a high temperature, which permanently hardens the clay as chemical changes occur.

Group VII

Characteristic physical properties

Group VII consists of reactive non-metals, all of which exist as diatomic molecules, X_2. Their properties are summarised in Table 10.5.

Table 10.5

Element	Appearance	Boiling point/K	$E^\ominus$: $X_2 + 2e^- \rightleftharpoons 2X^-$/V	Electronegativity
Fluorine	Pale yellow gas	85	+2.87	4.0
Chlorine	Yellow-green gas	238	+1.36	3.0
Bromine	Dark red liquid	332	+1.07	2.8
Iodine	Shiny dark grey solid	457	+0.54	2.5

The Group VII elements (halogens) are volatile non-metals. They exist as diatomic molecules that attract each other by van der Waals forces. The larger the halogen, the greater the van der Waals forces and hence the higher the boiling point.

There is a decrease in reactivity on moving down the group. This is due, in part, to the increase in atomic radius because the incoming electron has to go into a shell further away from an increasingly shielded nucleus. Nonetheless, the elements are still too reactive to occur uncombined, unlike some metals and other non-metals such as carbon. The electronegativity and the $E^\ominus$ values show that these elements are oxidising agents with reactivity that decreases down the group.

Important chemical reactions

Hydrogen halides

One of the most important groups of compounds of the Group VII elements is the hydrogen halides, HX. The hydrogen halides are prepared in different ways depending on the oxidising power of the halogen concerned. Hydrogen chloride can be prepared by heating sodium chloride with concentrated sulfuric acid:

$$NaCl(s) + H_2SO_4(l) \rightarrow NaHSO_4(s) + HCl(g)$$

However, neither hydrogen bromide nor hydrogen iodide can be prepared by this method because they would be oxidised by the sulfuric acid.

The only acid that can be used to prepare all three hydrogen halides is phosphoric(V) acid:

$$NaX(s) + H_3PO_4(s) \rightarrow NaH_2PO_4(s) + HX(g)$$

Hydrogen fluoride is much harder to produce in a pure state since fluorine is such a strong oxidising agent. It will even oxidise water, giving a mixture of hydrogen fluoride and oxygen:

$$2F_2(g) + 2H_2O(l) \rightarrow 4HF(g) + O_2(g)$$

You have to be able to compare the bond energies of the hydrogen halides (Table 10.6) and use these data to explain their relative thermal stabilities.

Table 10.6

Element	Bond energy/kJ mol^{-1}			
	Fluoride	Chloride	Bromide	Iodide
H	568	432	366	298
C	467	346	290	228

The bond energies of fluorine with hydrogen and carbon are significantly higher than those of the other Group VII elements. This means that the formation of covalent fluorides is usually strongly exothermic since this means breaking F–F bonds and making E–F bonds (where E is the element concerned).

The H–X bond energies decrease steadily down the group, making it easier to break the H–X bond. So, for example, plunging a red-hot wire into a halogen halide has the following effects:

- HI decomposes
- HBr shows some evidence of decomposition
- HC*l* is unaffected

Reactions of chlorine with sodium hydroxide

Chlorine reacts differently with sodium hydroxide depending on the temperature and concentration of the alkali.

With cold, dilute sodium hydroxide solution the following reaction takes place:

$$Cl_2(aq) + 2OH^-(aq) \rightarrow Cl^-(aq) + ClO^-(aq) + H_2O(l)$$

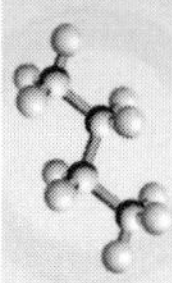

In this reaction, the element chlorine (oxidation number 0) has been converted into chloride ions (oxidation number –1) and chlorate(I) ions (oxidation number +1).

With hot, concentrated sodium hydroxide solution the following reaction takes place:

$$3Cl_2(aq) + 6OH^-(aq) \rightarrow 5Cl^-(aq) + ClO_3^-(aq) + 3H_2O(l)$$

In this case, the element chlorine has been converted into chloride ions (oxidation number –1) and chlorate(V) ions (oxidation number +5).

These are both examples of **disproportionation** reactions in which an element is both oxidised and reduced. The reason for the difference in the two reactions is the instability of the chlorate(I) ion, in which the chlorine disproportionates again at higher temperatures:

$$3ClO^-(aq) \rightarrow 2Cl^-(aq) + ClO_3^-(aq)$$

Reactions of the halide ions, other than fluoride

Testing for halide ions

The characteristic test for halide ions in solution is to add silver nitrate solution followed by aqueous ammonia (Table 10.7). You have probably carried out this test a number of times in practical sessions.

Table 10.7

Halide ion	Reaction with $Ag^+(aq)$	Subsequent reaction with $NH_3(aq)$
Chloride	White precipitate is formed	Dissolves to form colourless solution
Bromide	Cream precipitate is formed	Only dissolves in concentrated ammonia
Iodide	Yellow precipitate is formed	Insoluble in ammonia

$$Ag^+(aq) + X^-(aq) \rightleftharpoons Ag^+X^-(s)$$

$$Ag^+(aq) + 2NH_3(aq) \rightleftharpoons [Ag(NH_3)_2]^+(aq)$$

The equilibrium in the first equation lies to the right so it can be regarded as a forward reaction. However, the equilibrium in the second equation also lies to the right and, therefore, removes silver ions from the remaining solution, causing the silver halide to dissolve. This is true for chloride and bromide ions. However, silver iodide is so insoluble that even concentrated ammonia solution is unable to reverse the process.

Reactions with other halogens

The halide ions in solution react with halogens higher up the group, being oxidised to their respective halogen. Aqueous chlorine reacts with both bromide and iodide ions, liberating bromine and iodine respectively:

$$2Br^-(aq) + Cl_2(aq) \rightarrow Br_2(aq) + 2Cl^-(aq)$$

Aqueous bromine liberates iodine from iodide ions.

Reactions with concentrated sulfuric acid

The reactions of the halide ions with concentrated sulfuric acid can also be used as a test (Table 10.8).

Table 10.8

Ion present	Observations
Chloride	Steamy acidic fumes (of HC*l*)
Bromide	Steamy acidic fumes (of HBr) mixed with brown bromine vapour
Iodide	Some steamy fumes (of HI) but lots of purple iodine vapour

Although the hydrogen halide is formed in each case, hydrogen bromide and hydrogen iodide are oxidised by the sulfuric acid, as shown in the equations:

$$NaCl(s) + H_2SO_4(l) \rightarrow NaHSO_4(s) + HCl(g)$$

$$2HBr(g) + H_2SO_4(l) \rightarrow SO_2(g) + 2H_2O(l) + Br_2(l)$$

$$8HI(g) + H_2SO_4(l) \rightarrow H_2S(g) + 4H_2O(l) + 4I_2(s)$$

Note that one molecule of sulfuric acid oxidises two molecules of hydrogen bromide, but eight molecules of hydrogen iodide. This shows how much easier it is to oxidise iodide ions, I^-, than it is to oxidise bromide ions, Br^-.

Economic importance of halogens

You have already seen in Chapter 6 how chlorine is manufactured from brine using a diaphragm cell. Chlorine, and to a lesser extent the other halogens, have a number of important economic and industrial uses.

Chlorine in water purification

Chlorine is used to kill bacteria and so sterilise water for domestic supplies in many parts of the world; it is also used in some swimming pools. The ability of chlorine to destroy bacteria is a result of its powerful oxidising power, which disrupts the chemistry of bacterial cells. However, traces in our drinking water are insufficient to do us any harm and the benefits of water chlorination far outweigh the risk. The chlorine may be supplied as the gas, or added as solid sodium chlorate(I).

Manufacture of bleach

Sodium hydroxide and chlorine can be chemically combined to make the bleach, sodium chlorate(I), NaC*l*O. This is used in some domestic cleaning agents. It chemically cleans materials such as washbasins and toilets, and 'kills' germs.

Manufacture of PVC

PVC, or poly(chloroethene), is a tough, useful plastic and a good electrical insulator. It is much tougher than poly(ethene) and very hard wearing with good heat stability. It is used for covering electrical wiring and plugs. It is also replacing metals for use as gas and water pipes, and replacing wood in window frames. It is made by reacting chlorine with ethene, heating the product and then polymerising the chloroethene formed (see Chapter 19).

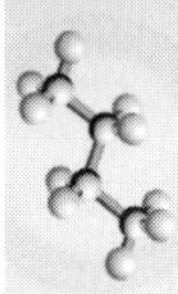

Halogenated hydrocarbons

Many hydrocarbons react with halogens and a number of the products are commercially important. You have already seen how chlorine is important in making the polymer, PVC. It is also important in making solvents such as trichloroethane, used in dry-cleaning, and carbon tetrachloride used in organic chemistry. It is used to produce antiseptics and disinfectants such as TCP and Dettol, and has been used to make CFCs (chlorofluorocarbons), which, in the past, were used as refrigerants and aerosol propellants. This use has now been reduced dramatically because CFCs break down in the upper atmosphere, and the chlorine atoms catalyse the destruction of ozone, which is important for absorbing ultraviolet radiation.

11 The transition elements

This chapter is needed for the A2 examination. The only part relevant to AS is the use of iron as a catalyst in the Haber process and vanadium pentoxide as a catalyst in the Contact process, which are dealt with in Chapter 12.

Physical properties

Electronic structures of atoms and ions

Although you can probably identify the block of elements that make up the transition elements in the periodic table (Figure 11.1), you need to know *why* these elements are different from other metals, and what makes them special.

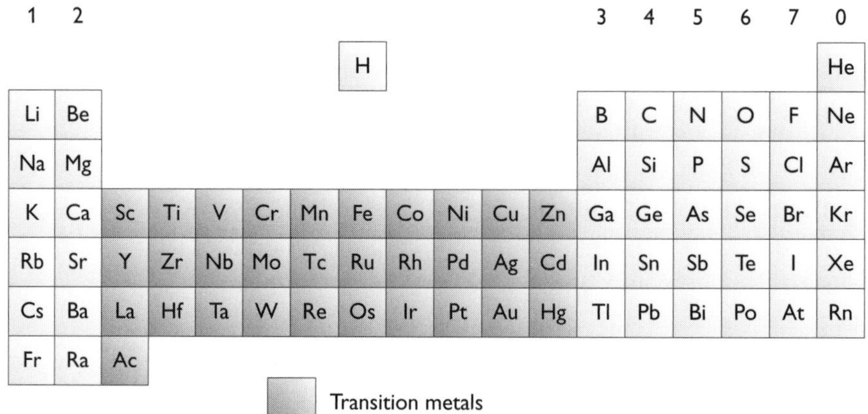

Figure 11.1 The periodic table

The block of elements shown represents the filling of the d-electron orbitals for each of the 3d-, 4d- and 5d-orbitals. You only really need to know details of the 3d elements — scandium, Sc, to zinc, Zn. A transition element is defined as a metal that forms one or more stable ions with incompletely filled d-orbitals. We can best see this by looking at the electron configuration of the elements and ions of this group of elements (Figure 11.2).

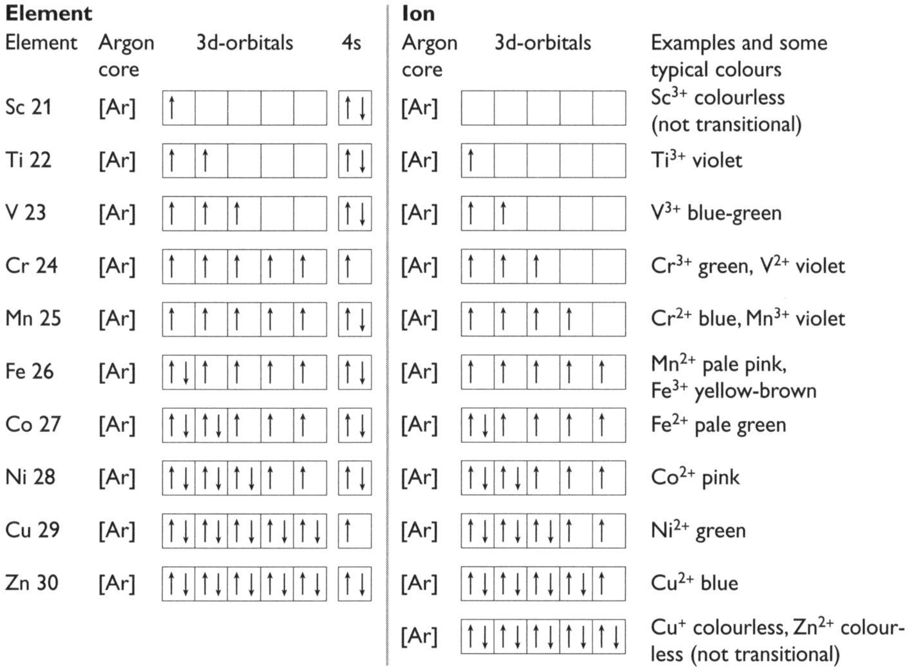

Element									**Ion**						
Element	Argon core	3d-orbitals					4s		Argon core	3d-orbitals					Examples and some typical colours
Sc 21	[Ar]	↑					↑↓		[Ar]						Sc^{3+} colourless (not transitional)
Ti 22	[Ar]	↑	↑				↑↓		[Ar]	↑					Ti^{3+} violet
V 23	[Ar]	↑	↑	↑			↑↓		[Ar]	↑	↑				V^{3+} blue-green
Cr 24	[Ar]	↑	↑	↑	↑	↑	↑		[Ar]	↑	↑	↑			Cr^{3+} green, V^{2+} violet
Mn 25	[Ar]	↑	↑	↑	↑	↑	↑↓		[Ar]	↑	↑	↑	↑		Cr^{2+} blue, Mn^{3+} violet
Fe 26	[Ar]	↑↓	↑	↑	↑	↑	↑↓		[Ar]	↑	↑	↑	↑	↑	Mn^{2+} pale pink, Fe^{3+} yellow-brown
Co 27	[Ar]	↑↓	↑↓	↑	↑	↑	↑↓		[Ar]	↑↓	↑	↑	↑	↑	Fe^{2+} pale green
Ni 28	[Ar]	↑↓	↑↓	↑↓	↑	↑	↑↓		[Ar]	↑↓	↑↓	↑	↑	↑	Co^{2+} pink
Cu 29	[Ar]	↑↓	↑↓	↑↓	↑↓	↑↓	↑		[Ar]	↑↓	↑↓	↑↓	↑	↑	Ni^{2+} green
Zn 30	[Ar]	↑↓	↑↓	↑↓	↑↓	↑↓	↑↓		[Ar]	↑↓	↑↓	↑↓	↑↓	↑	Cu^{2+} blue
									[Ar]	↑↓	↑↓	↑↓	↑↓	↑↓	Cu^{+} colourless, Zn^{2+} colourless (not transitional)

Figure 11.2 Electron configuration of the elements and some ions of transition metals

Table 11.1

Property	Sc	Ti	V	Cr	Mn	Fe	Co	Ni	Cu	Zn
Melting point/°C	1541	1668	1910	1857	1246	1538	1495	1455	1083	420
Density/ $g\,cm^{-3}$	2.99	4.54	6.11	7.19	7.33	7.87	8.90	8.90	8.92	7.13
Atomic radius/pm	161	145	132	125	124	124	125	125	128	133
M^{2+} ionic radius/pm	n/a	90	88	84	80	76	74	72	69	74
M^{3+} ionic radius/pm	81	76	74	69	66	64	63	62	n/a	n/a
Common oxidation states	+3	+2,3,4	+2,3, 4,5	+2,3,6	+2,3, 4,6,7	+2,3,6	+2,3	+2,+3	+1,2	+2
Outer electron configuration	$3d^14s^2$	$3d^24s^2$	$3d^34s^2$	$3d^54s^1$	$3d^54s^2$	$3d^64s^2$	$3d^74s^2$	$3d^84s^2$	$3d^{10}4s^1$	$3d^{10}4s^2$
1st ionisation energy/ $kJ\,mol^{-1}$	632	661	648	653	716	762	757	736	745	908

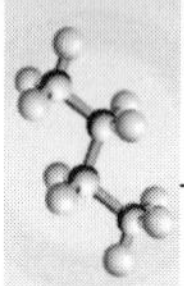

Table 11.1 shows some of the common properties of the elements scandium to zinc. If you look first at the electron configurations of the elements, you can see that all except copper and zinc have incompletely filled d-orbitals.

Next, considering the common oxidation states, you can see that Sc^{3+} ions have no d-electrons, and that Zn^{2+} ions have full d-orbitals. Based on this evidence there is a case for saying that scandium is a transition element, but that no case can be made for zinc.

On examining the common stable ions of the transition elements we can see that, with the exception of scandium, all the elements lose their two 4s-electrons to form a +2 ion. Where elements possess two to five 3d-electrons the loss of these, together with the 4s-electrons, gives the highest oxidation state (the ion with the greatest positive charge). Note that there are no d^4 or d^9 arrangements for chromium or copper. This is because half-full and completely full d-orbitals are favoured energetically.

In the examination you might be asked to state the electron configuration of a given transition element and suggest its stable ions. You can do this from the periodic table:

- first, use the proton (atomic) number to work out the total number of electrons
- then, put the electrons into orbitals, bearing in mind the exceptions mentioned above

Look at the rows in Table 11.1 that give data on atomic radii, ionic radii and first ionisation energies. You can see that across the transition elements there is a relatively small change in each of these properties.

Other physical properties

You are also expected to be able to contrast the properties of the transition elements with those of the s-block metal, calcium (Table 11.2).

Table 11.2

Property	Ca	Sc	Ti	V	Cr	Mn	Fe	Co	Ni	Cu
Melting point/°C	839	1541	1668	1910	1857	1246	1538	1495	1455	1083
Density/g cm^{-3}	1.55	2.99	4.54	6.11	7.19	7.33	7.87	8.90	8.90	8.92
Atomic radius/pm	197	161	145	132	125	124	124	125	125	128
M^{2+} ionic radius/pm	106	n/a	90	88	84	80	76	74	72	69
1st ionisation energy/kJ mol^{-1}	590	632	661	648	653	716	762	757	736	745

You can see from the melting points and densities that the transition elements are much more similar to each other than they are to calcium. This is also borne out by the atomic and M^{2+} ionic radii and, to a lesser extent, by the first ionisation energies. The syllabus also refers to conductivity, but comparisons are harder to make here since all metals are much better electrical conductors than semi-conductors or non-metals.

Chemical properties

Oxidation states

As you can see from Table 11.1, most of the transition elements form more than one ion or oxidation state. You can see the details more easily in Figure 11.3. The reason that these other oxidation states exist is that there is not a large energy barrier to the removal of subsequent electrons.

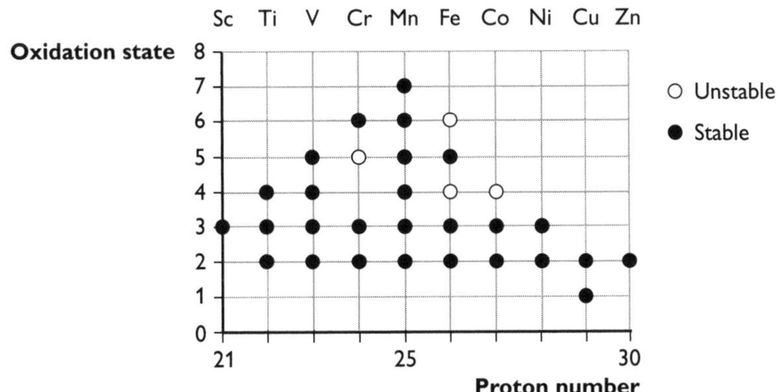

Figure 11.3 Oxidation states in transition metals

You can see an example of this if you compare the first four ionisation energies of calcium with those of chromium and manganese (Table 11.3).

Table 11.3

Element	Proton number	1st I.E./ kJ mol^{-1}	2nd I.E./ kJ mol^{-1}	3rd I.E./ kJ mol^{-1}	4th I.E./ kJ mol^{-1}
Calcium	20	590	1150	4940	6480
Chromium	24	653	1590	2990	4770
Manganese	25	716	1510	3250	5190

It is clear from the data in Table 11.3, that to remove the third electron from calcium requires about as much energy as removing the fourth from chromium or manganese. However, it is not as simple as just comparing ionisation energies. Having ionised the metal, it has to react to form a compound. There are two key enthalpies to consider — the lattice enthalpy (if a solid is being formed) and the enthalpies of hydration of the ions (if an aqueous solution is being formed).

The more highly charged the ion, the more electrons have to be removed and the more ionisation energy has to be provided. Off-setting this, however, the more highly charged the ion, the more energy is released either as lattice enthalpy or as the hydration enthalpy of the metal ion.

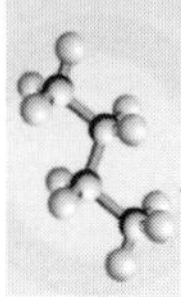

Try this yourself

(27) The graph below was obtained for one of the first row transition metals. Explain the important points on the graph and hence identify the metal.

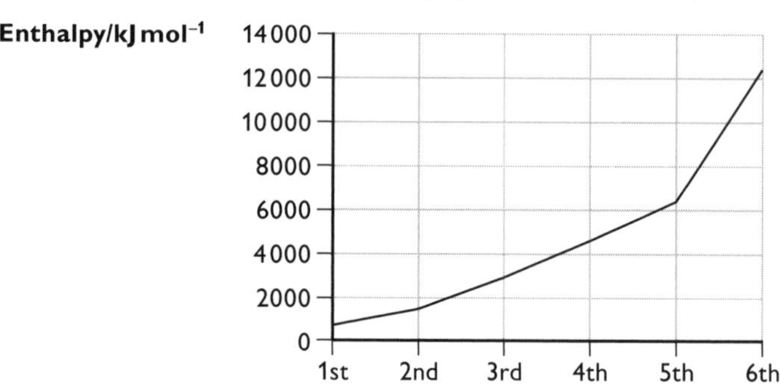

Redox systems

There are three important redox systems that you need to know for the theory and also for the practical syllabus: Fe^{3+}/Fe^{2+}, MnO_4^-/Mn^{2+} and $Cr_2O_7^{2-}/Cr^{3+}$. We have already looked at some simple redox processes in Chapter 6.

The reaction between acidified manganate(VII)ions and iron(II)ions

This reaction is used to estimate iron(II) ions quantitatively. It is self-indicating. On the addition of a standard solution of potassium manganate(VII) to an iron(II) solution, decolorisation of manganate(VII) occurs as the almost colourless manganese(II) ion (a *very* pale pink) is formed. The end point is the first permanent pale pink with one drop excess of the manganate(VII). The presence of dilute sulfuric acid prevents the formation of a manganese(IV) oxide precipitate and ensures the reduction of the manganate(VII) ion to the manganese(II) ion.

The two half-equations are:

$$MnO_4^-(aq) + 8H^+(aq) + 5e^- \rightleftharpoons Mn^{2+}(aq) + 4H_2O(l) \qquad E^\ominus = +1.52\,V$$

$$Fe^{3+}(aq) + e^- \rightleftharpoons Fe^{2+}(aq) \qquad E^\ominus = +0.77\,V$$

Since each iron(II) ion supplies one electron, each manganate(VII) ion can oxidise five iron(II) ions. Thus, the overall equation is:

$$MnO_4^-(aq) + 8H^+(aq) + 5Fe^{2+}(aq) \rightarrow Mn^{2+}(aq) + 5Fe^{3+}(aq) + 4H_2O(l)$$

The reaction between acidified dichromate(VI) ions and iron(II) ions

As with potassium manganate(VII), a standard solution of potassium dichromate(VI) can be used to estimate iron(II) ions in solution quantitatively. In this case, however, a redox indicator must be used to detect the end point.

The indicator changes colour when oxidised to another form, but only after the iron is oxidised, i.e. it is not as easily oxidised as Fe^{2+}— the $E^\ominus$ of the indicator is more positive than that of Fe^{2+} but less than that for the dichromate(VI) ion. Hence, it is oxidised by the dichromate to show the end point.

The two half-equations are:

$Cr_2O_7^{2-}(aq) + 14H^+(aq) + 6e^- \rightleftharpoons 2Cr^{3+}(aq) + 7H_2O(l)$ $E^\ominus = +1.33\,V$

$Fe^{3+}(aq) + e^- \rightleftharpoons Fe^{2+}(aq)$ $E^\ominus = +0.77\,V$

Since each iron(II) ion supplies one electron, each dichromate(VI) ion can oxidise six iron(II) ions. Thus the overall equation is:

$$Cr_2O_7^{2-}(aq) + 14H^+(aq) + 6Fe^{2+}(aq) \rightarrow 2Cr^{3+}(aq) + 6Fe^{3+}(aq) + 7H_2O(l)$$

Other redox reactions

The transition metals take part in a range of redox reactions, some of which could form part of the titrimetric work you are asked to carry out in the practical paper. You might also be asked to predict whether or not a given reaction will take place based on $E^\ominus$ data. If you are not sure how to do this, re-read Chapter 6.

The important things to remember are:

- Metals react by electron loss (oxidation state increases) to form a positive cation (e.g. sodium ion, Na^+) — so, as the electron loss potential increases, the metal reactivity of the metal increases.
- Non-metallic elements react by electron gain (oxidation state decrease) to form a single covalent bond (e.g. HC*l*) or the negative anion (e.g. chloride ion C*l*⁻ in NaC*l*) — so, as the electron accepting power decreases, so does the reactivity of the element.
- For a reaction to be feasible, the E_{cell} value must be *positive* — so if you calculate it to be negative, the reverse reaction will be the feasible one.

Try this yourself

(28) Work out the overall equation for the reaction of manganate(VII) ions with hydrogen peroxide in acid solution.

The geometry of the d-orbitals

In Chapter 3 we looked at the shape and symmetry of the s- and p-orbitals, which can hold two and six electrons respectively. In considering the transition elements you now need to understand the shape and symmetry of the d-orbitals, which can hold up to ten electrons.

The d-orbitals can be divided into two groups, as shown in Figure 11.4.

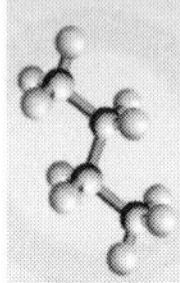

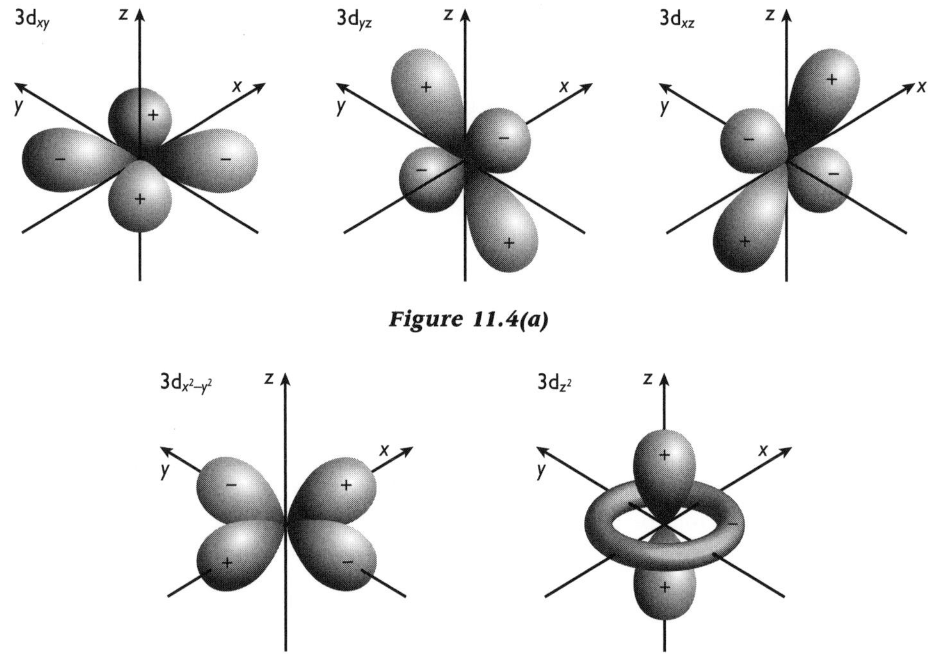

Figure 11.4(a)

Figure 11.4(b)

Each of the first group is similar to a pair of p-orbitals at right angles to one another, and as you can see the lobes lie between the *xy*, *yz* and *zx* axes.

The second group is different in that the lobes lie on the *xy* and *z* axes. You do *not* need to know *why* they are these shapes, only the shapes and symmetry of the five orbitals.

The shape and symmetry of these orbitals are important when transition metal cations react to form complexes. The metal cations do this by interacting with **ligands**. A ligand is an atom, ion or molecule that can act as an electron pair donor, and which usually forms a dative covalent or 'coordinate' bond with the central metal ion.

Imagine electron-rich ligands approaching a transition metal ion with electrons in its d-orbitals. There will be some repulsion, raising the energy of some d-orbitals more than that of others.

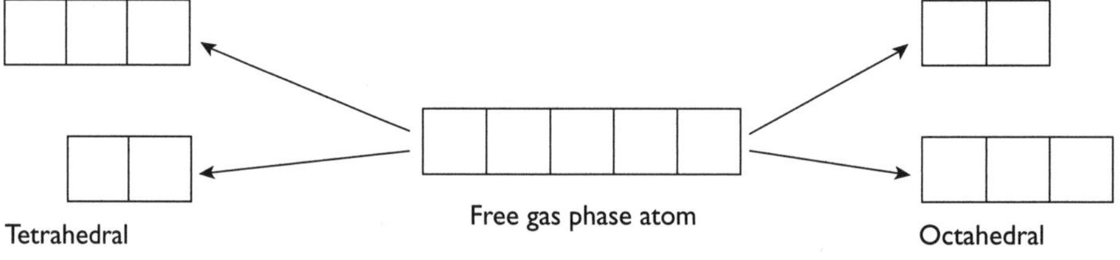

Figure 11.5 Splitting of d-orbitals in transition metal complexes

Figure 11.5 shows that for the formation of a tetrahedral complex, three orbitals have higher energy ($3d_{xy}$, $3d_{yz}$ and $3d_{zx}$), whereas for an octahedral complex the reverse is true ($3d_{x^2-y^2}$ and $3d_{z^2}$). This is because in octahedral complexes the ligands approach the central metal ion along the axes, and there is repulsion between the electrons on the ligands and those in the $3d_{x^2-y^2}$ and $3d_{z^2}$ orbitals. In tetrahedral complexes the four ligands approach the central metal ion between the axes and now there is repulsion between the electrons on the ligands and those in the $3d_{xy}$, $3d_{yz}$ and $3d_{zx}$ orbitals.

Formation of complexes

As well as forming simple compounds such as oxides and salts, one of the characteristic properties of transition metals is their ability to form complex ions, many of which have a distinctive colour. The reason for this colour is the absorption of light in different parts of the electromagnetic spectrum. This absorption occurs because of the movement of an electron from a lower energy d-orbital to one of higher energy (see Figure 11.5). In the case of the aqueous copper(II) ion the transition is shown in Figure 11.6.

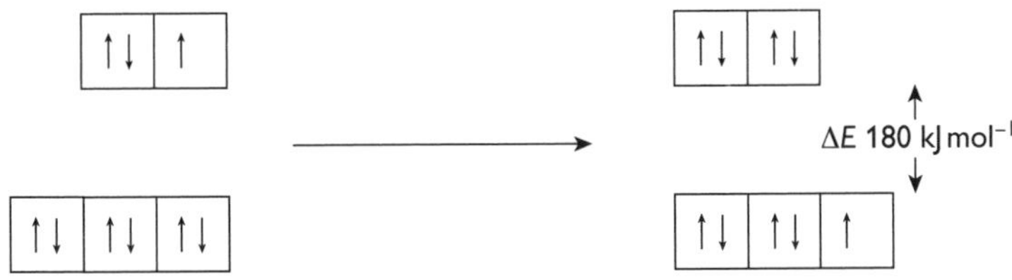

Figure 11.6 Promotion of an electron in copper(II)

The promotion of the electron requires 180 kJ mol^{-1}, and the frequency corresponding to this lies in the red-orange region of the spectrum. As a result, these colours are absorbed leaving yellow, green, blue and purple to be transmitted, resulting in the familiar pale blue colour of aqueous Cu(II).

The colour of a complex depends on the energy gap in the d-orbitals, which is a result of two factors:

- the nature of the metal and its oxidation state
- the nature of the ligand

You are familiar with the colours of the two common oxidation states of iron, Fe^{2+} and Fe^{3+} in aqueous solution: Fe^{2+} is pale green and Fe^{3+} is yellow-brown.

When copper ions are dissolved in water, they form the complex ion $[Cu(H_2O)_6]^{2+}$, which is pale blue. On adding aqueous ammonia to this solution until the ammonia is present in an excess, the solution turns a deep blue:

$$[Cu(H_2O)_6]^{2+} + 4NH_3 \rightleftharpoons [Cu(NH_3)_4(H_2O)_2]^{2+} + 4H_2O$$

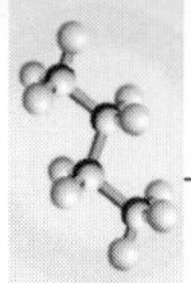

The deepening blue is a sign of a greater energy gap between the two sets of orbitals (Figure 11.7).

Figure 11.7 Energy gap in copper(II) complexes

Other ligands can have a greater or lesser energy gap and some indication of the order is given below.

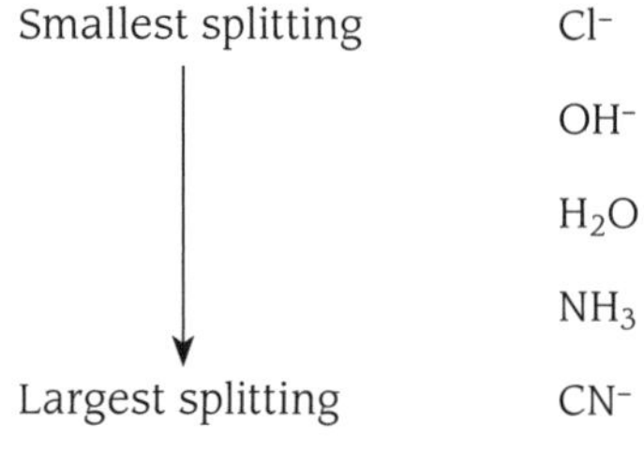

Try this yourself

(29) Three octahedral complexes of chromium are coloured blue, green and purple. Each complex contains one type of ligand — ammonia, water or hydroxide ions.

Identify the formula of each complex and match it to the correct colour, explaining your reasoning.

Complex stability and ligand exchange

The different colour changes seen on adding different reagents to aqueous solutions that contain transition metal ions occur because some complexes are more stable than others. Remember that most aqueous reactions are equilibria and that using a high concentration of a reagent will change the position of equilibrium. However, this only becomes significant if the stability constant for the complex has a reasonable magnitude. The stability constant is the term used for the equilibrium constant for any change.

We have already seen how, in the case of aqueous copper(II) ions and ammonia, the pale blue of the hexaaquacopper(II) ion is replaced by the much darker blue tetraamminediaquacopper(II) ion. You can probably think of other examples.

12 Nitrogen and sulfur

This chapter is needed for the AS examination.

Nitrogen

The lack of reactivity

We know that nitrogen is an unreactive gas because it is mixed with reactive oxygen in Earth's atmosphere and reacts very little. The reason for this lack of reactivity is the very strong N≡N bond in the molecule. The two nitrogen atoms share three electron pairs, which form a triple covalent bond, and each atom retains a lone pair of electrons. This is shown in Figure 12.1. The bond energy for nitrogen is 946 kJ mol^{-1} compared with the bond energy in fluorine, F_2, which is 158 kJ mol^{-1}.

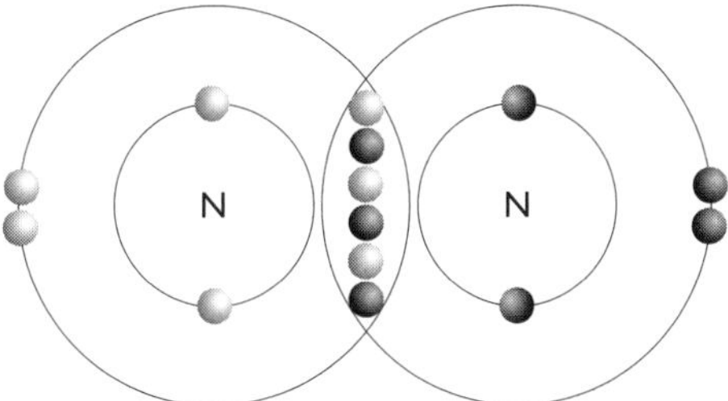

Figure 12.1 Structure of nitrogen

Although nitrogen is unreactive, it does react under the right conditions. For example, burning magnesium reacts with the nitrogen in air to form magnesium nitride:

$$3Mg(s) + N_2(g) \rightarrow Mg_3N_2(s)$$

At high temperatures nitrogen will react with oxygen to form oxides of nitrogen, for example:

$$N_2(g) + O_2(g) \rightleftharpoons 2NO(g)$$

In car engines (and in thunderstorms) nitrogen combines with oxygen to produce a mixture of oxides of nitrogen often referred to as NO_x.

Nitrogen also reacts with hydrogen to form ammonia:

$$N_2(g) + 3H_2(g) \rightleftharpoons 2NH_3(g)$$

This is the basic reaction in the Haber process for the manufacture of ammonia. This is covered in Chapter 7 and in more detail later in this section.

Finally, the roots of some plants of the pea and bean family have nodules that contain bacteria able to 'fix' nitrogen chemically. The bacteria convert the nitrogen into ammonium ions, which the plants can use to make proteins.

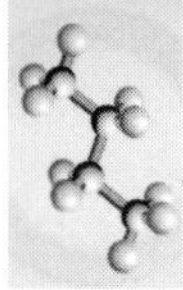

Ammonia

Ammonia, NH_3, is an alkaline gas. You might expect it to have a trigonal planar shape, but it is pyramidal with a lone pair of electrons occupying the apex, as is shown in Figure 12.2.

Figure 12.2 Structure of ammonia

Remember that the lone pair takes up more space than a pair of bonding electrons. In the case of ammonia this reduces the H—N—H bond angles to about 107°. The high electronegativity (3.0) of nitrogen means that ammonia can form hydrogen bonds, and as a result it is very soluble in water.

Ammonia is also a base, accepting a proton to form the ammonium ion, NH_4^+:

$$NH_3(g) + H^+(aq) \rightleftharpoons NH_4^+(aq)$$

This ion is tetrahedral, the proton forming a coordinate (dative) bond with the nitrogen atom using the lone pair (Figure 12.3).

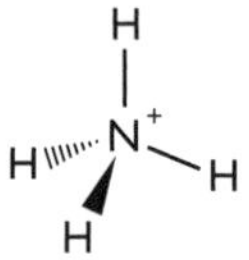

Figure 12.3 Structure of the ammonium ion

In the laboratory, ammonia is easily displaced from ammonium compounds by warming with a strong base such as sodium hydroxide:

$$NH_4^+(aq) + OH^-(aq) \rightarrow NH_3(g) + H_2O(l)$$

The Haber process

We have already looked at some aspects of the Haber process in Chapters 7 and 8. You may wonder why you have to study this manufacturing process. Ammonia is one of the most important bulk chemicals manufactured, mainly because of its use in the production of fertilisers. In 2008, more than 150 million tonnes of ammonia were produced worldwide.

The original Haber process was developed almost 100 years ago and still forms the basis of ammonia production, although the pressures and temperatures have been refined to give maximum production at an optimum price. Most recent research has focused on improving the iron catalyst. Both osmium and ruthenium are more effective as catalysts, but both are more expensive than iron. Therefore, the catalyst used depends on the market price of ammonia.

For the examination you need to be able to state the essential operating conditions for an ammonia plant (Figure 12.4), and to interpret these in terms of reaction kinetics and the position of equilibrium.

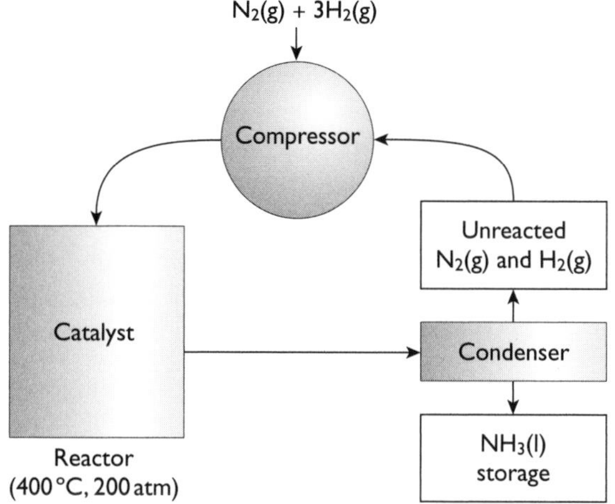

Figure 12.4 The Haber process

The set of conditions chosen are a compromise:

- pressure — around 20 000 kPa (or 200 atm)
- temperature — around 670 K
- catalyst — iron (which has a lifetime of around 5 years)

As well as its use in fertiliser manufacture (mainly as the sulfate or nitrate), ammonia can be oxidised using a platinum catalyst to form nitrogen monoxide:

$$4NH_3(g) + 5O_2(g) \rightarrow 4NO(g) + 6H_2O(l) \quad \Delta H^\circ = -909\,kJ\,mol^{-1}$$

This exothermic reaction is the starting point for the manufacture of nitric acid. The hot gas is cooled, reacted with more oxygen to form nitrogen dioxide, and then dissolved in water to form nitric acid:

$$2NO(g) + O_2(g) \rightleftharpoons 2NO_2(g)$$

$$2NO_2(g) \rightleftharpoons N_2O_4(g)$$

$$3N_2O_4(g) + 2H_2O(l) \rightarrow 4HNO_3(aq)$$

Nitrogen compounds and pollution

Nitrates

Nitrogen compounds, particularly nitrates, are used to make fertilisers. These have had significant benefits in increasing crop yields around the world. However, if excessive amounts of fertilisers are used they can get washed into streams and rivers. This can have two effects:

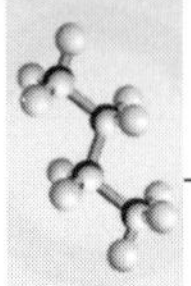

- The nitrates can get into drinking water supplies, from which they are difficult to remove. They can affect the ability of babies under 6 months to carry oxygen in the bloodstream.
- They can increase the growth of aquatic vegetation, which then decays reducing oxygen levels in streams and rivers, and affecting other forms of aquatic life.

Nitrogen oxides

The combustion of motor fuels produces temperatures high enough to form oxides of nitrogen, NO_x, in exhaust gases. These gases in air at street level are known to contribute to respiratory problems. They also have a pollution effect in the upper atmosphere where they catalyse the oxidation of sulfur dioxide. The main source of sulfur dioxide in the atmosphere is the combustion of fuels (mainly coal and oil) that contain sulfur or its compounds. Some of the sulfur dioxide is removed from the waste gases emitted by major users of these fuels, such as power stations.

The exact mechanism is uncertain, but the following is a possibility:

$$2NO_2(g) + H_2O(g) \rightarrow HNO_2(l) + HNO_3(l)$$

$$SO_2(aq) + HNO_3(l) \rightarrow NOHSO_4(l)$$

$$NOHSO_4(l) + HNO_2(l) \rightarrow H_2SO_4(l) + NO_2(g) + NO(g)$$

$$SO_2(aq) + 2HNO_2(l) \rightarrow H_2SO_4(l) + 2NO(g)$$

The other form of atmospheric pollution caused by nitrogen oxides is the formation of ozone, O_3, in the presence of unburnt hydrocarbons from vehicles. Ozone formed at street level can cause respiratory problems. However, once nitrogen oxides escape into the upper atmosphere they cause the reverse problem, removing ozone and reducing the ability of ozone to protect the Earth's surface from harmful ultraviolet radiation.

The emission of nitrogen oxides from vehicles has been greatly reduced since the introduction of catalytic converters. These also reduce the emission of carbon monoxide and unburnt hydrocarbons (see also Chapter 8).

$$2NO_x(g) \rightarrow xO_2(g) + N_2(g)$$

$$2CO(g) + O_2(g) \rightarrow 2CO_2(g)$$

$$C_nH_{2n+2}(g) + [(3n + 1)/2]O_2(g) \rightarrow nCO_2(g) + (n + 1)H_2O(g)$$

Sulfur

The Contact process

Sulfuric acid is an extremely important industrial chemical that is used in the production of a huge range of manufactured goods. Figure 12.5 shows some of the major uses of sulfuric acid. In 2008, world production of sulfuric acid was more than 200 million tonnes.

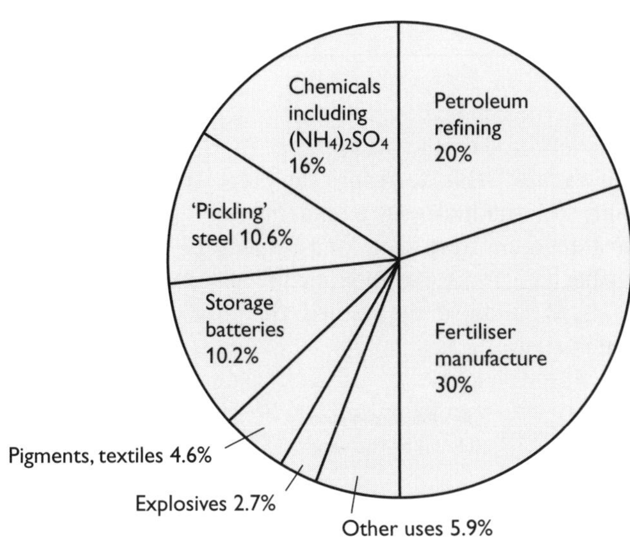

Figure 12.5 Uses of sulfuric acid

For the examination you need to be able to state the essential operating conditions for the manufacture of sulfuric acid by the Contact process (Figure 12.6), and to interpret these in terms of reaction kinetics and the position of equilibrium.

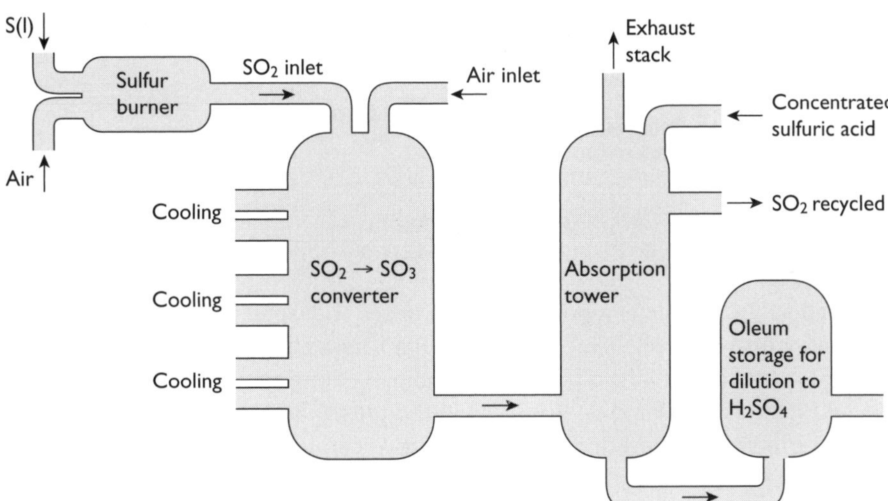

Figure 12.6 The Contact process

The conditions chosen are a compromise:

- pressure — around 100–200 kPa (or 1–2 atm)
- temperature — around 700 K
- catalyst — vanadium(V) oxide (although some plants now use ruthenium which is more expensive)

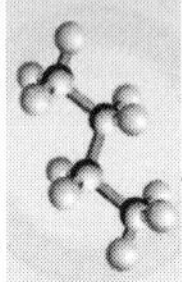

Sulfur dioxide

Acid rain

We have already seen that the combustion of sulfur-containing fuels releases sulfur dioxide into the atmosphere, and that in the presence of oxides of nitrogen, this can be converted into sulfuric acid. This decreases the pH of rain. Acid rain can harm plants and animals directly, and indirectly by making lakes acidic. It also releases toxic metals such as aluminium from soils, and below pH 4.5 no fish are likely to survive. This has further effects up the food chain. Acid rain also increases the erosion of limestone-based buildings and statues. The production and some effects of acid rain are shown in Figure 12.7.

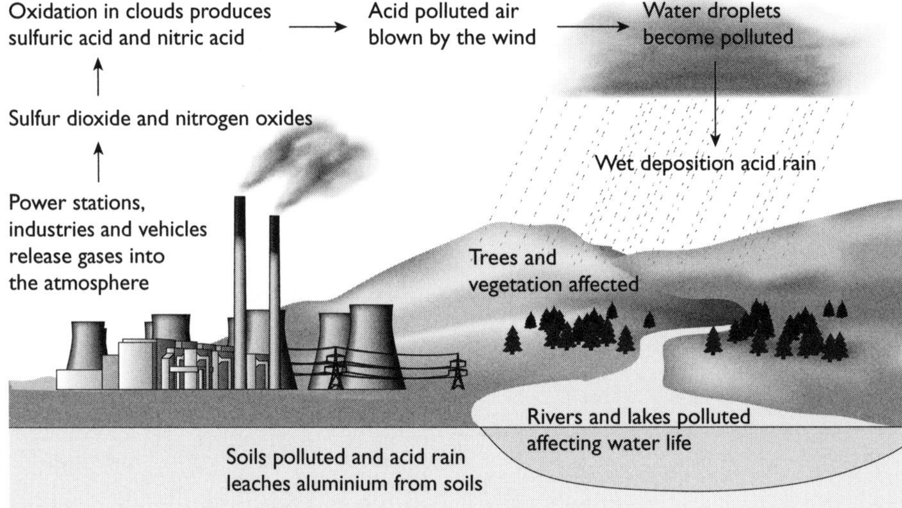

Figure 12.7 Acid rain

Food preservation

Although sulfur dioxide is a pollutant that causes widespread damage to the environment, it also has beneficial uses, particularly as a preservative for foodstuffs. It has the advantage of being a reducing agent (preventing spoilage by oxidation) and has anti-microbial properties. It is used in the preservation of fruit products (such as dried fruits and juices) and in winemaking. It can be a problem for people who suffer from asthma.

13 Introduction to organic chemistry

Most of this chapter is needed for the AS examination. The aspects that are needed only for A2 are clearly indicated.

Formulae

It is important to understand and know when to use the different ways of representing organic molecules. Read the examples carefully so that you are in no doubt.

Molecular formula

A molecular formula summarises the numbers and types of atom present in a molecule. The functional group is shown separately from the hydrocarbon chain — for example, C_2H_5OH rather than C_2H_6O.

Structural formula

A structural formula requires the minimum detail to provide an unambiguous structure for a compound. For example, $CH_3CH_2CH_2OH$ is acceptable for propan-1-ol whereas C_3H_7OH is not.

Displayed formula

A displayed formula shows the correct positioning of the atoms and the bonds between them. For example ethanoic acid, which has the structural formula CH_3CO_2H, has the displayed formula:

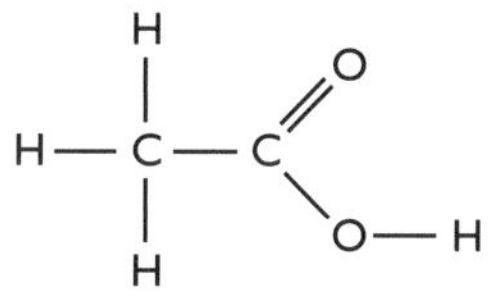

You may be asked for 'partially displayed formulae'. This means that you have to show the positions of atoms and the bonds between them at the site of a reaction.

Skeletal formula

This is a simplified representation of an organic molecule that concentrates on the carbon 'backbone' of a molecule, together with any functional groups. Bonds to hydrogen atoms are *not* normally shown, unless they form part of a functional group. The skeletal formula for butan-2-ol is shown below:

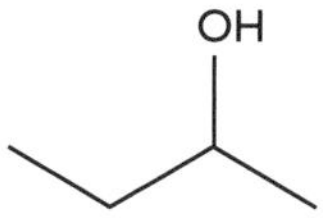

For more complex structures, where the three-dimensional structure of the molecule may be important, a 'partial-skeletal' formula may be used. This shows the geometry of key bonds to hydrogen (and other) atoms in the molecule, as shown below:

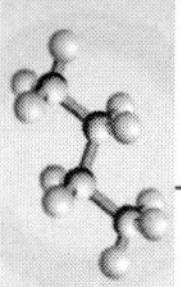

Three-dimensional structures

There are times when you need to be able to show the three-dimensional structures of relatively simple molecules, — for example, showing a pair of optical isomers. The convention of using a solid wedge to represent a bond coming 'out of the paper' and a dashed line for one going 'behind the paper' is used, as shown below:

Mirror plane

Try this yourself

(30) What is the name of the compound shown in the diagram immediately above?

Names and functional groups

It is important that you know:

- how to name hydrocarbon chains
- how to name the functional groups in organic chemistry
- how to indicate the positions of functional groups in the molecule

There are some simple rules that will help you.

The hydrocarbon chain

The key thing to remember is that the prefix indicates the number of carbon atoms present in the main chain (Table 13.1).

Table 13.1

Number of carbon atoms	Prefix used
1	meth-
2	eth-
3	prop-
4	but-
5	pent-
6	hex-
7	hept-
8	oct-

Hydrocarbon molecules do not have only straight chains, they can be branched. For a branched molecule, look at the number of carbon atoms in the branch and count the number of the carbon to which the branch is joined (remember to count from the end that gives the lower number). Some examples are shown below. Remember to use the *longest continuous* carbon chain as the basic hydrocarbon.

So, $CH_3CH_2CH_2CH_2CH_2CH_3$ is hexane.

$CH_3CH_2CHCH_2CH_2CH_3$ (with CH_2CH_3 attached to the third carbon) or $CH_3CH_2CH(CH_2CH_3)CH_2CH_2CH_3$

are both 3-ethylhexane (an ethyl group on carbon 3)

$CH_3CH_2CCH_2CH_3$ (with CH_3 above and CH_3 below the third carbon) or $CH_3CH_2C(CH_3)_2CH_2CH_3$

are both 3,3-dimethylpentane (two methyl groups on carbon 3).

Alkanes

Alkanes are a family of hydrocarbons that contain only C—C single bonds and C—H bonds. Alkanes are relatively unreactive, except to combustion and they form the major fuels that we use. All of the examples given above are alkanes.

Alkenes

Alkenes are a family of hydrocarbons that have a reactive functional group, the C=C double bond. The double bond makes alkenes more reactive and they are important organic compounds.

In alkenes, it is the position of the C=C double bond that is indicated. So, $CH_3CH_2CH{=}CHCH_2CH_3$ is called hex-3-ene and $(CH_3)_2C{=}CHCH_2CH_3$ is called 2-methylpent-2-ene (the double bond is in the second possible position and there is a methyl group branching from carbon atom 2).

Arenes

Arenes are a family of hydrocarbons that contain a benzene ring made up of six carbons.

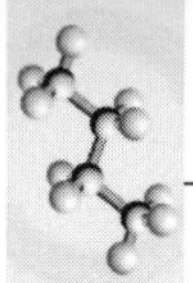

Benzene

In a benzene ring the carbon atoms are numbered clockwise from the uppermost atom. You only need to use this numbering system if there is more than one group attached to the ring.

Tip Arenes are important for the A2 examination. At AS you need to be able to recognise the benzene ring, even though you do not need to know anything about benzene or its compounds.

Other functional groups

Some functional groups are shown in Table 13.2.

Table 13.2

Name of compound	Formula of group
Halogenoalkane/arene	–Hal
Alcohol/phenol	–OH
Aldehyde	–CHO
Ketone	–C=O
Carboxylic acid	$-CO_2H$
Ester	$-CO_2R$ (where R is a hydrocarbon group)
Acyl chloride	–CO*Cl*
Amine	$-NH_2$
Nitrile	–CN
Amide	$-CONH_2$

Naming compounds is not too difficult. Some examples are shown in Table 13.3.

Table 13.3

Formula	Name of compound
$CH_3CH_2CH_2Br$	1-bromopropane
CH_3CH_2OH	Ethanol
$CH_3CH_2CH_2CHO$	Butanal
$(CH_3)_2C{=}O$	Propanone
$CH_3CH_2CO_2H$	Propanoic acid
$CH_3CH_2CO_2CH_3$	Methyl propanoate
CH_3COCl	Ethanoyl chloride
$CH_3CH_2CH_2CH_2NH_2$	1-aminobutane
$CH_3CH_2CH_2CN$	Propanenitrile
CH_3CONH_2	Ethanamide

Try this yourself

(31) Name the following compounds:

(a) $CH_3CH_2CH_2CH_2Cl$

(b) $CH_3CHOHCH_3$

(c) $CH_3CH_2CH_2CH_2CH_2CO_2H$

(d) $CH_3CH(CH_3)CH_2CH_2NH_2$

(32) Draw structural formulae for the following compounds:

(a) 2-bromobutane

(b) propanamide

(c) methanal

(d) methyl ethanoate

Organic reactions

It is important to be able to remember, and in some cases to predict, what types of reaction a compound containing a particular functional group will take part in. To do this you need to be aware of both the nature of the functional group and of the possible reactions a given reagent may permit.

Fission of bonds

Organic compounds are held together by covalent bonds, so you no longer need to worry about ionic reactions (except in a few rare cases). In organic molecules, a given bond can split in two ways. In **homolytic** fission one electron goes to each fragment:

H
H C Cl → H C + Cl
H

In **heterolytic** fission both electrons go to one fragment and none to the other:

H
H C Cl → [H C]$^+$ + [Cl]$^-$
H

Free radical reactions

Free radicals are usually highly reactive species consisting of an atom or fragment of a molecule with an unpaired electron. In equations, it is usual to show the unpaired electron as a 'dot'. These free radicals may be formed by the action of ultraviolet light (such as in the upper atmosphere) or by the breakdown of a very unstable organic compound.

In general, free radical reactions take place in three distinct steps — initiation, propagation and termination. The reaction you will study is the reaction of methane with chlorine in the presence of ultraviolet light.

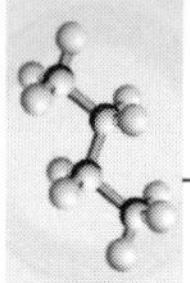

Step 1: initiation Free radicals are formed by the breaking of a bond by homolytic fission:

$$Cl\text{–}Cl \xrightarrow{UV} Cl^{\bullet} + Cl^{\bullet}$$

Step 2: propagation The free radicals formed begin a chain reaction in which as each free radical is used another is formed:

$$H\text{–}CH_3 + Cl^{\bullet} \rightarrow CH_3^{\bullet} + HCl$$

and $CH_3^{\bullet} + Cl\text{–}Cl \rightarrow CH_3Cl + Cl^{\bullet}$

Step 3: termination These are reactions in which free radicals combine and hence end that part of the chain reaction:

$$Cl^{\bullet} + Cl^{\bullet} \rightarrow Cl_2$$

and $CH_3^{\bullet} + Cl^{\bullet} \rightarrow CH_3Cl$

and $CH_3^{\bullet} + CH_3^{\bullet} \rightarrow CH_3CH_3$

Nucleophilic and electrophilic reactions

Nucleophilic reagents 'like' positive charges and electrophilic reagents 'like' areas of electron density.

Tip Here is a way to remember this — 'nucleo like nucleus, nuclei are positive; electro like electrons, which are negative'.

It follows that the reagents themselves are the opposite of what they seek.

Nucleophiles

Nucleophiles include halide ions, hydroxyl ions, cyanide ions and molecules containing lone pairs, such as water and ammonia, or even ethanol. A typical nucleophilic substitution reaction is:

$$CH_3CH_2CH_2Br + OH^- \rightarrow CH_3CH_2CH_2OH + Br^-$$

Electrophiles

Electrophiles are electron-deficient species, generally positively charged ions such as H^+, Cl^+, Br^+, I^+, NO_2^+, CH_3^+, CH_3CO^+. An example of an electrophilic addition reaction is:

$$CH_3CH{=}CH_2 \rightarrow HBr \rightarrow CH_3CHBr\ CH_3$$

Other types of reaction

Addition

Addition refers to an increase in saturation, in other words adding a molecule to a C=C double bond as in the above reaction. The molecule achieves this by interacting with the π-electrons in the double bond (see the structure of ethene on page 126).

Substitution

Substitution refers to the replacement of one group in an organic molecule by another as in nucleophilic substitution on page 140. Both nucleophiles and electrophiles can take part in substitution reactions.

Elimination

An elimination reaction involves the removal of atoms from two adjacent carbon atoms to leave a double bond. It is the reverse of the electrophilic addition reaction on page 135.

Hydrolysis

This is a reaction, usually in aqueous media, between one organic molecule and water or acid or alkali which leads to the formation of at least two products. Two examples are:

$$(CH_3)_3C{-}Cl + 2H_2O \rightarrow (CH_3)_3C{-}OH + H_3O^+ + Cl^-$$

$$CH_3CO_2CH_3 + NaOH \rightarrow CH_3CO_2^-Na^+ + CH_3OH$$

Oxidation

In general, this refers to the oxidation of a C—OH group to form a C=O group in an aldehyde, ketone or carboxylic acid. Such oxidations are often brought about by warming the organic compound with acidified potassium dichromate(VI). This can produce complicated equations, and in organic reactions it is permissible to show the oxidising agent as [O]:

$$CH_3CH_2OH + [O] \rightarrow CH_3CHO + H_2O$$

Reduction

This is the opposite of oxidation and, in general, applies to compounds containing a C=O group. Reduction may be brought about by several reducing agents including tin and dilute hydrochloric acid, sodium in ethanol and lithium aluminium hydride. In organic reduction reactions, the reducing agent is usually represented by [H], as shown in the example below:

$$CH_3CO_2H + 4[H] \rightarrow CH_3CH_2OH + H_2O$$

Shapes of molecules

For AS, you need to know the shapes of ethane and ethene. For A2, you also need to know the shape of benzene. Alongside this, you need to be able to work out the shapes of related molecules. The basics of this were covered in Chapter 3, so here are some reminders.

Ethane

An ethane molecule is formed by electrons in hydrogen 1s-orbitals overlapping with electrons in $2sp^3$-orbitals on the carbon atoms to form molecular orbitals in which

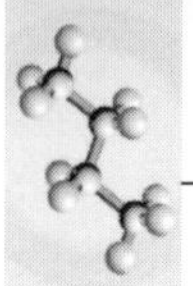

the hydrogen nuclei are embedded. A single C—C σ-bond, formed by the overlapping of one sp^3-orbital from each carbon atom, joins the two ends together, but there is no restriction on rotation so the two ends of the molecule can spin relatively freely.

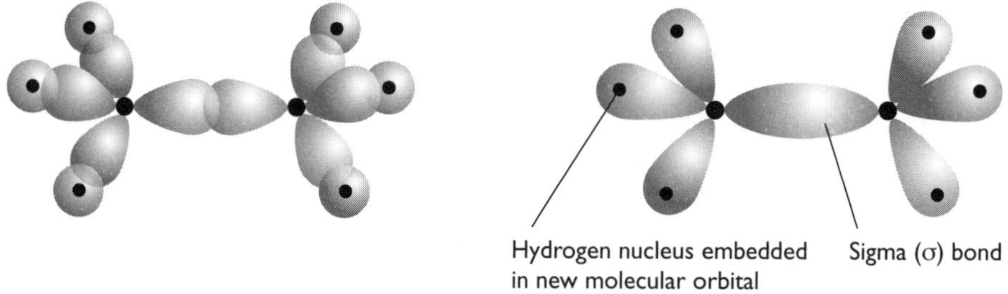

Ethene

In ethene, the carbon atoms form $2sp^2$-hybrid orbitals for three of the electrons, leaving one electron in a 2p-orbital. An ethene molecule is formed by the overlap of two sp^2-orbitals on each carbon with two hydrogen 1s-orbitals, with the third sp^2-orbital on each carbon atom overlapping to form a σ-bond. The p-orbitals interact to form an additional π-bond which prevents the rotation of the ends of the molecule about the σ-bond:

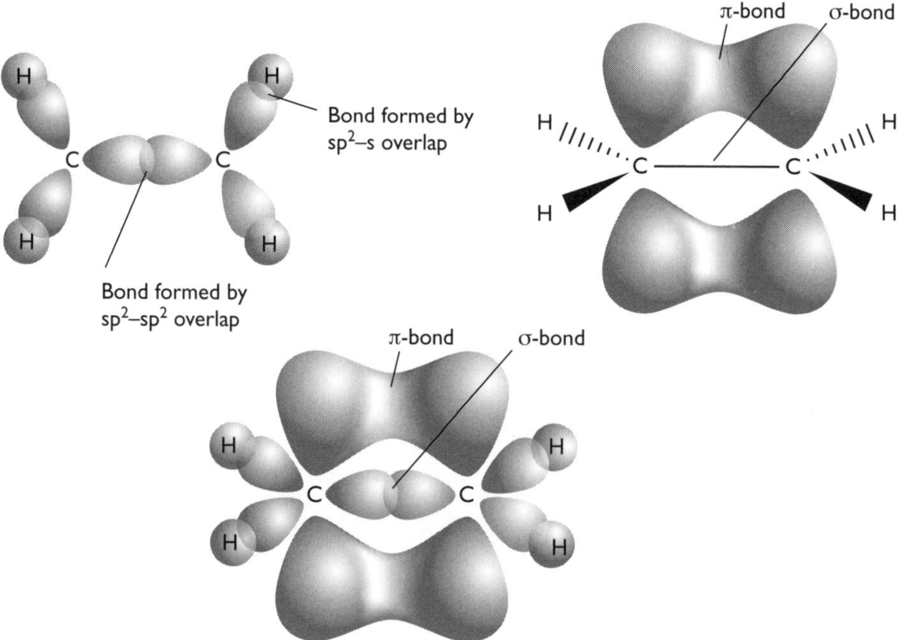

Benzene

A molecule of benzene has six carbons arranged in a hexagonal ring. These carbon atoms have hybridised orbitals in the same manner as ethene. However, in benzene, only one $2sp^2$-orbital is bonded to hydrogen, with the other two sp^2-orbitals being bonded to adjacent carbons. The 2p-orbitals interact around the ring producing a π-electron 'cloud' above and below the plane of the ring.

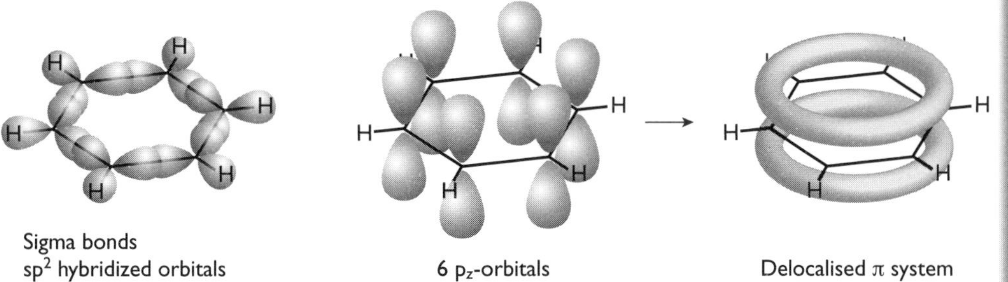

Sigma bonds
sp^2 hybridized orbitals

6 p_z-orbitals

Delocalised π system

Isomerism

Isomers are compounds that have the same molecular formula (same chemical composition) but different structural formulae. You need to know about three main types of isomerism — structural, geometrical and optical.

Structural isomerism

Chain isomerism

In **chain isomerism**, the isomers arise due to branching of the carbon chain. So C_4H_{10} can have both a straight chain form and a branched chain form:

$CH_3CH_2CH_2CH_3$ — butane

$CH_3CH(CH_3)CH_3$ — 2-methylpropane

The compounds have different names based on the system described on page 121. The branched chain is three carbons long and has a methyl group on the second carbon atom.

Position isomerism

In **position isomerism**, the carbon chain is fixed, but the position of substituent groups can vary. The alcohols propan-1-ol and propan-2-ol show this:

$CH_3CH_2CH_2OH$ — Propan-1-ol

$CH_3CH(OH)CH_3$ — Propan-2-ol

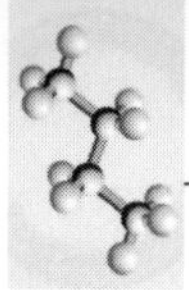

Changing the position of a group can affect how easily it reacts, as well as its physical properties, such as boiling point. Propan-1-ol has a boiling point of 97 °C whereas propan-2-ol has a boiling point of 82 °C.

Functional group isomerism

In **functional group isomerism** the nature of the functional group within the molecule is different. This is important because it changes the chemical reactions that the molecule undergoes. The formula C_3H_6O can represent three molecules, each with a different functional group:

$CH_3—CH_2—CHO$ (C double-bonded to O, single-bonded to H)
Propanal

$(CH_3)_2C=O$
Propanone

$CH_2=CH—CH_2—OH$
2-propen-1-ol

Geometric or *cis-trans* isomerism

Geometric isomerism occurs where there is restricted rotation around a bond, such as in alkenes. It also needs two groups, one on each end of the double bond, such as in 1,2-dichloroethene:

ClHC=CHCl (both Cl on the same side)
cis-1,2-dichloroethene

ClHC=CHCl (Cl on opposite sides)
trans-1,2-dichloroethene

These two forms are different because the double bond prevents the rotation needed to make the two forms identical. Notice that there is another isomer, but this time it is a position isomer, rather than a *cis-trans* isomer.

$Cl_2C=CH_2$
1,1-dichloroethene

To summarise, in *cis-trans* isomerism:

- there is restricted rotation generally involving a carbon-carbon double bond
- there are two different groups on the left-hand end of the bond and two different groups on the right-hand end

Optical isomerism and chiral centres

The final type of isomerism you need to be able to recognise and explain is **optical isomerism**. This gets its name from the effect an optical isomer has on the plane of plane-polarised light. One isomer rotates the polarised light clockwise, and the other isomer rotates it an equal amount anti-clockwise. This occurs when there is a carbon with four different groups attached to it — this is called a **chiral** carbon. An example was shown on page 120 and another example, butan-2-ol, is shown below:

CH_2CH_3 | H_3C | C | H | OH

CH_3CH_2 | H | C | CH_3 | OH

Mirror plane

Notice that the two molecules are mirror-images, and that the central carbon atom has four different groups attached. When you draw these structures it is important to check that your diagrams make chemical 'sense'. Take the structure of 2-aminopropanoic acid for example. It is important to show the 'acid' group joined to the central carbon the right way round. (If not, you may be penalised in the exam.) Figure 13.1 shows what you need to do.

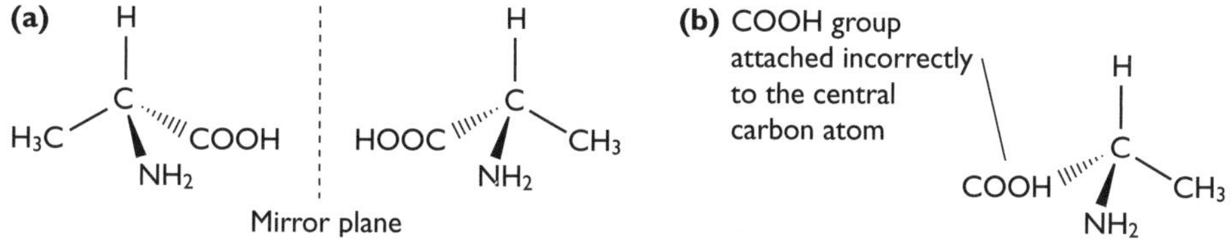

(a) H | H_3C | C | COOH | NH_2 H | HOOC | C | CH_3 | NH_2

Mirror plane

(b) COOH group attached incorrectly to the central carbon atom

H | COOH | C | CH_3 | NH_2

Figure 13.1(a) Optical isomers of 2-aminopropanoic acid (b) Incorrectly drawn isomer

Identifying chiral carbon atoms can be tricky, particularly in complex or skeletal molecular structures. You have to work out whether or not a given carbon atom has two (or more) identical groups attached to it.

Look at the skeletal structure below:

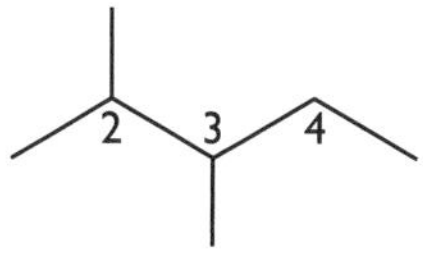

2 3 4

It contains three non-terminal carbon atoms numbered 2, 3 and 4.

Look at each of these in turn. Carbon 2 has two bonds attached to methyl groups, carbon 4 has two bonds attached to hydrogen atoms (not shown), and carbon 3 is attached to four different groups. Therefore, carbon 3 is chiral.

You may be asked to examine a complex molecule and state the number of chiral carbon atoms (or perhaps circle them). An example of such a molecule, the synthetic form of the hormone testosterone, is shown in Figure 13.2.

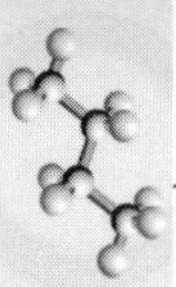

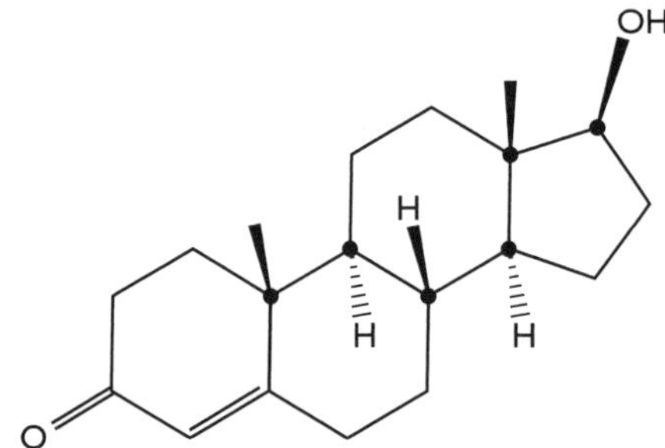

Figure 13.2 Synthetic testosterone

Can you identify the chiral carbon atoms in the structure? There are six, marked •.

Try this yourself

(33) A compound **P** has the formula C_4H_8O.

- **(a)** Draw a straight-chain structure for **P**.
- **(b)** Does **P** have functional group isomers? If so, draw examples.
- **(c)** How many isomers in total does the straight-chain form of **P** have, incorporating any of the functional groups in Table 13.2?

(34) The compound **Q** has the formula C_4H_7Br and contains a double bond.

- **(a)** How many non-cyclic isomers exist for **Q**?
- **(b)** Draw the *cis-trans* isomers of **Q**.

14 Hydrocarbons

Apart from the section on arenes, all this chapter is needed for the AS examination. Some of the content has been covered previously or you may have studied it at IGCSE.

Alkanes

Combustion

Due to their general lack of reactivity, the single most important use of alkanes is as fuels. You may already know about the importance of crude oil, and the 'cracking' of less useful fractions to form more useful products (this is covered on the next pages).

Ethane is used as an example of an alkane, noting that it has the formula C_2H_6 and that the general formula for alkanes is C_nH_{2n+2}.

Ethane reacts differently with oxygen depending on how much oxygen is available.

Plenty of oxygen: $2C_2H_6 + 7O_2 \rightarrow 4CO_2 + 6H_2O$

Less oxygen: $2C_2H_6 + 5O_2 \rightarrow 4CO + 6H_2O$

Restricted oxygen: $2C_2H_6 + 3O_2 \rightarrow 4C + 6H_2O$

Restricting the amount of oxygen reduces the amount by which the carbon in ethane is oxidised until it cannot be oxidised at all. The midway point produces poisonous carbon monoxide, which has been known to kill people using faulty gas heaters.

Crude oil as a source of hydrocarbons

The use of oil as a major fuel is less than 150 years old and results from the revolutions in land and air transport brought about by the development of the internal combustion engine. Crude oil is often talked about as if it is a mixture with fixed composition. However, it can vary enormously in the proportions of the various hydrocarbons that make up the mix. This variation affects how much processing the oil requires to yield useful products.

Not all sources of oil have high proportions of the hydrocarbons that are most in demand. However, chemists have developed ways of converting less useful hydrocarbons into more useful ones. The main process for achieving this is 'cracking' — a large molecule hydrocarbon of limited use is broken into smaller molecule hydrocarbons that are in greater demand.

Some of the most important processes in an oil refinery are shown in Figure 14.1.

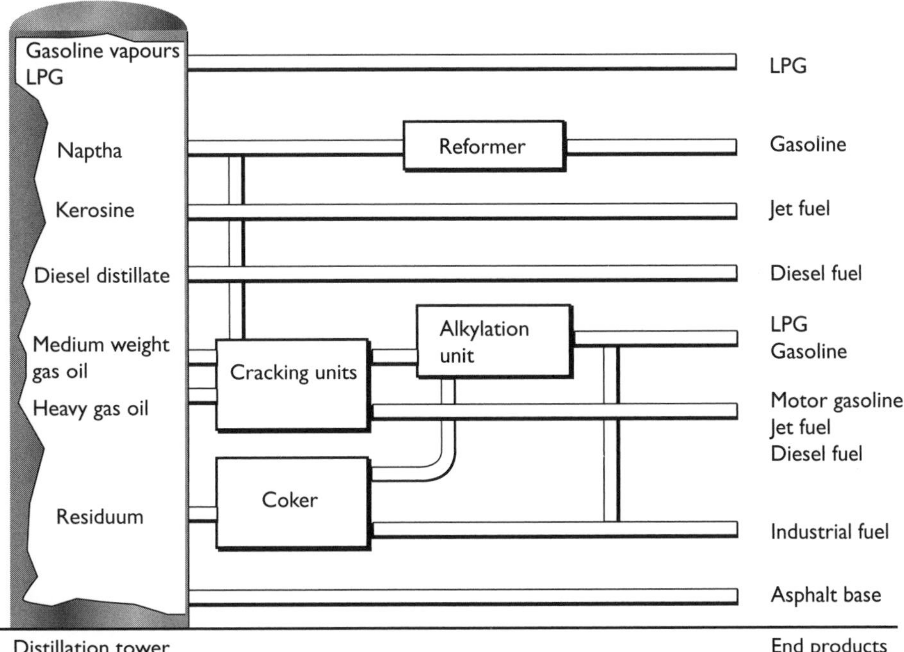

Figure 14.1 Processes in an oil refinery

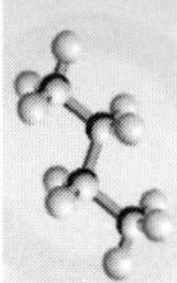

The cracking process involves using either high pressures and temperatures without a catalyst, or lower temperatures and pressures in the presence of a catalyst. The source of the large hydrocarbon molecules is often the 'naphtha' fraction or the 'gas oil' fraction from the fractional distillation of crude oil. Although these fractions are obtained as liquids, they have to be vaporised before cracking can occur.

There is no unique reaction in the cracking process. In general, a large hydrocarbon molecule produces one or more smaller alkane molecules and one or more alkene molecules, for example:

$$C_{15}H_{32} \rightarrow 2C_2H_4 + C_3H_6 + C_8H_{18}$$

$$C_{12}H_{26} \rightarrow C_3H_6 + C_9H_{20}$$

$$C_8H_{18} \rightarrow C_6H_{14} + C_2H_4$$

The alkanes produced are usually used for motor fuel — either petrol (gasoline) or diesel — while the alkenes formed are used in the polymer industry.

Substitution reactions

Alkanes react with difficulty with both chlorine and bromine. In order to react, alkanes need energy from ultraviolet light (sunlight) and, as you might expect, chlorine reacts more easily than bromine. Taking ethane as the example, the hydrogen atoms are replaced one at a time in substitution reactions:

$$C_2H_6 + Cl_2 \rightarrow C_2H_5Cl + HCl$$

$$C_2H_5Cl + Cl_2 \rightarrow C_2H_4Cl_2 + HCl$$

$C_2H_4Cl_2 + Cl_2 \rightarrow C_2H_3Cl_3 + HCl$ and so on

The mechanism for the equivalent reaction with methane was covered on page 124.

Alkenes

Ethene is used as an example of an alkene, noting that it has the formula C_2H_4 and that the general formula for alkenes is C_nH_{2n}.

Addition reactions

Because alkenes have a double bond, it is reasonable to expect addition reactions to be particularly important.

With hydrogen

Ethene reacts with hydrogen at a temperature of about 150 °C in the presence of finely-divided (powdered) nickel. The hydrogen adds across the double bond forming ethane:

$$CH_2{=}CH_2 + H_2 \rightarrow CH_3{-}CH_3$$

This is not a very useful reaction, but for larger alkenes — such as those found in vegetable oils — the addition of hydrogen across a double bond is more important. These oils are 'hardened', or turned into solid fats, by hydrogenation — a process that is necessary for the manufacture of margarine.

With steam

Water, in the form of steam, can be added across ethene's double bond to form ethanol. This is carried out industrially at a temperature of about 300°C and a pressure of about 60 atm in the presence of a phosphoric(V) acid catalyst.

For alkenes other than ethene, there is the possibility of adding the hydrogen to two different carbons. In propene, for example, the hydrogen can be added to either the end carbon or to the middle carbon, forming propan-2-ol and propan-1-ol respectively:

$$CH_3{-}CH{=}CH_2 + H_2O \rightarrow CH_3CHOHCH_3 \text{ or } CH_3CH_2CH_2OH$$

Markovnikoff's rule (worth learning) states that when a molecule of the form HX is added across a double bond, the hydrogen usually becomes attached to the carbon that is already attached to the most hydrogen atoms. This means that in the above reaction, propan-2-ol is favoured.

With hydrogen halides

If a gaseous alkene is bubbled through, or a liquid alkene is shaken with, either a concentrated aqueous solution of hydrogen bromide or hydrogen bromide dissolved in a non-polar solvent, the hydrogen bromide is added across the double bond.

The reaction is similar to the addition of water and follows Markovnikoff's rule:

$$CH_3{-}CH{=}CH_2 + HBr \rightarrow CH_3CHBrCH_3$$

With halogens

Ethene reacts with halogens by adding across the double bond. Therefore, with bromine at room temperature the reaction forms 1,2-dibromoethane:

$$CH_2{=}CH_2 + Br_2 \rightarrow CH_2BrCH_2Br$$

Chlorine, being more reactive, reacts faster than bromine. Iodine reacts more slowly.

The reaction above refers to ethene reacting with the pure halogen. Often, as in testing for alkenes using bromine water, a competing reaction can take place:

$$CH_2{=}CH_2 + Br_2 + H_2O \rightarrow CH_2BrCH_2OH + HBr$$

This compound is 1-bromo-2-hydroxyethane (or 2-bromethanol).

Oxidation reactions

Alkenes react with oxidising agents such as acidified manganate(VII) ions. The extent of the reaction, and hence the nature of the products, depends on the concentration of the oxidising agent and the temperature.

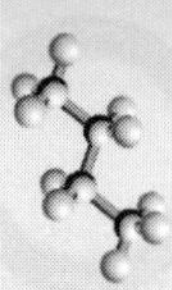

With cold, dilute, acidified manganate(VII) ions

Under these conditions ethene is oxidised to the diol 1,2-dihdroxyethane.

$$CH_2{=}CH_2 + [O] + H_2O \rightarrow CH_2OHCH_2OH$$

With hot, concentrated, acidified manganate(VII) ions

Acidified manganate(VII) is such a strong oxidising agent that in concentrated solution and with heat the carbon-to-carbon double bond of the alkene is ruptured. You may not think that this is a very useful reaction, but by looking at the products the position of a double bond in an unknown alkene can be determined.

The symbol **R** is used to represent a hydrocarbon group or a hydrogen atom. So, any alkene can be represented by the following formula:

$$(R_1)(R_3)C{=}C(R_2)(R_4)$$

When the acidified manganate(VII) ions oxidise the alkene, two C=O double bonds are formed:

$$(R_1)(R_3)C{=}C(R_2)(R_4) + 2[O] \rightarrow (R_1)(R_3)C{=}O + O{=}C(R_2)(R_4)$$

A compound that contains a C=O functional group is known as a **carbonyl compound**. A carbonyl compound with two hydrocarbon groups is called a **ketone**. If one of the **R** groups is hydrogen, the carbonyl compound formed is called an **aldehyde** and this can be oxidised further by acidified manganate(VII) to form a **carboxylic acid**:

$$(R_1)(H)C{=}C(R_2)(R_4) + 2[O] \rightarrow (R_1)(H)C{=}O + O{=}C(R_2)(R_4)$$

$$(R_1)(H)C{=}O + [O] \rightarrow (R_1)(HO)C{=}O$$

There is one further complication that occurs when there are no **R** groups at one end of the double bond. The carboxylic acid formed under those circumstances (methanoic acid) is itself oxidised by the acidified manganate(VII) ions to form carbon dioxide and water:

$$\text{HCHO (H}_2\text{C=O)} + [O] \rightarrow \text{HCOOH (H(HO)C=O)} + [O] \rightarrow CO_2 + H_2O$$

Here are some simple rules that might help you to work out the structure of the original alkene:

- Think about each end of the double bond separately.
- If there are two hydrocarbon groups at one end of the bond, that part of the molecule will give a ketone.
- If there is one hydrocarbon group and one hydrogen atom at one end of the bond, that part of the molecule will give a carboxylic acid.
- If there are two hydrogen atoms at one end of the bond, that part of the molecule will give carbon dioxide and water.
- Combine the information to work back to the structure of the original alkene.

Polymerisation

Carbon is one of the few elements to form rings and extended chains of atoms. Alkenes can join together to form long chains, or polymers. This does not apply only to hydrocarbon alkenes but also to substituted alkenes such as chloroethene ($CH_2{=}CHCl$) which is used to make PVC. Polymerisation is covered in Chapter 19.

Electrophilic addition

You saw in the previous chapter that halogens behave as electrophiles, and you know that alkenes have a concentration of electrons round the double bond. For the examination, you need to understand the mechanism of electrophilic addition, with the specific example of bromine reacting with ethene. The reaction takes place in two stages.

First, as the bromine molecule approaches the ethene molecule, the π-electrons in ethene induce a dipole on the bromine molecule. A bond is formed between the carbon and the bromine forming a positively charged species called a **carbocation**:

$$CH_2{=}CH_2 + {}^{\delta+}Br{-}Br^{\delta-} \longrightarrow CH_2Br{-}\overset{+}{C}H_2 + Br^-$$

Induced dipole

Second, the carbocation is attacked rapidly by the remaining Br^- ion to form the dibromide:

$$CH_2Br{-}\overset{+}{C}H_2 + Br^- \longrightarrow CH_2Br{-}CH_2Br$$

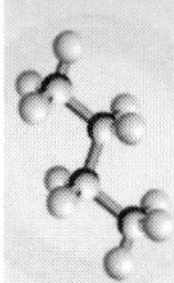

Try this yourself

(35) The reaction between bromine and ethene only occurs so 'cleanly' in a non-aqueous solvent. Explain what other product(s) might be formed if the bromine were dissolved in water.

Arenes

You need to know about two arenes — benzene and methylbenzene:

Benzene Methylbenzene

Although you might not think these molecules are very different, the presence of a side chain in methylbenzene means that it is able to undergo an additional set of reactions compared to benzene.

Substitution reactions

Benzene

Benzene reacts at room temperature with chlorine or bromine in the presence of a catalyst. One of the hydrogen atoms in the ring is replaced by a chlorine or bromine atom. A typical catalyst is the aluminium halide of the halogen being substituted, or iron (which reacts with the halogen to form the iron(III) halide which then acts as the catalyst):

$$C_6H_6 + Cl_2 \rightarrow C_6H_5Cl + HCl$$

$$C_6H_6 + Br_2 \rightarrow C_6H_5Br + HBr$$

In the presence of ultraviolet light (or sunlight), but without a catalyst, benzene undergoes addition reactions with both chlorine and bromine, with six halogen atoms added to the ring:

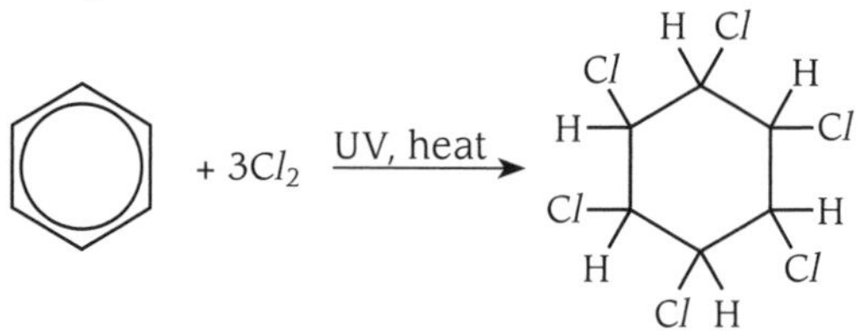

+ $3Cl_2$ $\xrightarrow{\text{UV, heat}}$

H Cl, Cl, H, H, Cl, Cl, H, H, Cl, Cl H

As you might expect, the reaction is faster with chlorine than with bromine.

Methylbenzene

With methylbenzene there are two different types of substitution, depending on whether a ring hydrogen or a methyl hydrogen is substituted.

As with benzene, substitution of a ring hydrogen occurs at room temperature in the presence of an aluminium halide or iron catalyst. There is an additional complication of where the halogen atom goes in relation to the methyl group. Methyl groups direct further substitution to the 2- or 4- positions in the ring (the 1-position is that occupied by the methyl group). This reaction with either chlorine or bromine under these conditions results in the formation of a mixture of 2-halo- and 4-halomethylbenzene:

$$C_6H_5CH_3 + Cl_2 \xrightarrow{AlCl_3\ cat} C_6H_4(CH_3)Cl \text{ (2-position)} + HCl$$

$$C_6H_5CH_3 + Cl_2 \xrightarrow{AlCl_3\ cat} C_6H_4(CH_3)Cl \text{ (4-position)} + HCl$$

When boiling methylbenzene is reacted with chlorine or bromine in the presence of ultraviolet light, the methyl hydrogen atoms are substituted. Provided sufficient halogen is present, all three hydrogen atoms are eventually substituted:

$$C_6H_5CH_3 + Cl_2 \xrightarrow{Heat} C_6H_5CH_2Cl + HCl$$

$$C_6H_5CH_2Cl + Cl_2 \xrightarrow{Heat} C_6H_5CHCl_2 + HCl$$

$$C_6H_5CHCl_2 + Cl_2 \xrightarrow{Heat} C_6H_5CCl_3 + HCl$$

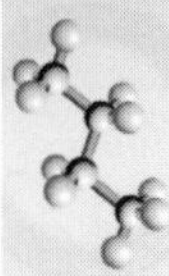

Nitration

This is the only reaction of arenes for which you need to know the mechanism.

Benzene

When benzene is treated with a mixture of concentrated nitric acid and concentrated sulfuric acid at a temperature lower than 50 °C, yellow nitrobenzene is gradually formed. The sulfuric acid acts as a catalyst.

$$C_6H_6 + HNO_3 \rightarrow C_6H_5NO_2 + H_2O$$

At higher temperatures, or with prolonged reaction even at 50 °C, further nitration occurs with a second nitro group being substituted into the ring. The second nitro group goes into the 3-position:

NO_2 (nitrobenzene) $+ HNO_3 \xrightarrow{\text{Reflux}}$ 1,3-dinitrobenzene (NO_2, NO_2) $+ H_2O$

Compare this with the methyl group in methylbenzene (p.137). It *is* possible to get a further nitro group in the 5-position, but the presence of a nitro group 'deactivates' the benzene ring, making it much less likely to react.

The mechanism for the mononitration of benzene is an example of electrophilic substitution. The nitrating mixture of concentrated nitric and concentrated sulfuric acids produces the electrophile — the nitronium ion, NO_2^+:

$$HNO_3 + 2H_2SO_4 \rightarrow NO_2^+ + 2HSO_4^- + H_3O^+$$

The NO_2^+ ion approaches the delocalised electrons in benzene and two of these form a bond with the positive charge now spread over the rest of the atoms in the ring:

benzene + $\overset{+}{NO_2}$ ⟶ intermediate (H, NO_2, +)

The HSO_4^- ion produced in the nitrating mixture now removes a hydrogen atom, re-forming the sulfuric acid catalyst:

HSO_4^- + intermediate (H, NO_2, +) ⟶ nitrobenzene (NO_2) $+ H_2SO_4$

Methylbenzene

In nitration, methylbenzene reacts about 25 times faster than benzene. This means that a lower temperature (around 30 °C rather than 50 °C) has to be used to prevent more than one nitro group being substituted. Apart from that, the reaction is the same and the same nitrating mixture of concentrated sulfuric and nitric acids is used.

As with the halogens, a mixture of the 2- and 4-nitro substituted arenes is formed:

$$C_6H_5CH_3 + HNO_3 \xrightarrow{30°C} \text{2-}NO_2C_6H_4CH_3 + H_2O$$

$$C_6H_5CH_3 + HNO_3 \xrightarrow{30°C} \text{4-}NO_2C_6H_4CH_3 + H_2O$$

Side-chain oxidation

This applies only to methylbenzene (and other arenes with alkyl side chains). Alkyl groups in alkanes are usually fairly unreactive towards oxidising agents. However, when attached to a benzene ring they are relatively easily oxidised. Heating methylbenzene (or any alkylbenzene) with alkaline potassium manganate(VII) solution, followed by acidification with dilute sulfuric acid, gives benzoic acid:

$$C_6H_5CH_3 \xrightarrow[\text{(2) Dilute } H_2SO_4]{\text{(1) Alkaline } KMnO_4} C_6H_5COOH$$

15 Halogen and hydroxy compounds

Most of this chapter, apart from the reactivity of chlorobenzene and the last section on hydroxyl compounds, is needed for the AS examination.

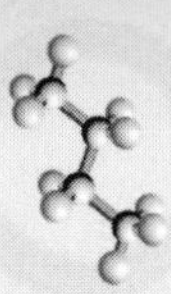

Halogen derivatives

Halogenoalkanes behave differently depending on which other groups are attached to the carbon that the halogen is attached to (you will see this trend with other functional groups).

- If there are only hydrogen atoms attached to the carbon, it is a **primary** (1°) halogenoalkane, for example CH_3CH_2Br.
- If there is one alkyl group attached as well as the halogen, it is a **secondary** (2°) halogenoalkane, for example $(CH_3)_2CHBr$
- If there are *no* hydrogen atoms, only alkyl groups and the halogen, it is a **tertiary** (3°) halogenoalkane, for example $(CH_3)_3CBr$

The C—Hal bond is polar due to the difference in electronegativity between the carbon atom and the halogen atom. Except when bonded to iodine, the carbon atom is relatively positive, making it susceptible to nucleophilic attack by lone pairs of electrons or negative ions.

Nucleophilic substitution

Halogenoalkanes undergo a number of nucleophilic substitution reactions. The syllabus requires you to know about three of these using bromoethane as a starting compound. However, you should be able to recognise this type of reaction with different halogenoalkanes and different nucleophilic reagents.

Hydrolysis

When bromoethane, a primary halogenoalkane, is heated under reflux with sodium hydroxide in a solvent of aqueous ethanol, the bromine is substituted by the hydroxyl group and ethanol is formed:

$$CH_3CH_2Br + OH^- \rightarrow CH_3CH_2OH + Br^-$$

You also need to know the mechanism for this reaction, which can be represented in two ways. The first way is as follows:

$$CH_3\text{—}\overset{\delta+}{C}H_2\text{—}\overset{\delta-}{Br} \xrightarrow{\quad} CH_3\text{—}CH_2\text{—}OH + :Br^-$$
$$:\bar{O}H$$

The reaction is described as S_N2, because there are two reactants in the rate-determining or slow step.

Tip Think of S_N2 as 'Substitution Nucleophilic, 2 reactants'.

The other way of representing this reaction is to show it as a two-stage process:

$$HO^{-} + CH_3CH_2Br\ (C^{\delta+}—Br^{\delta-}) \longrightarrow [HO\cdots C(CH_3)(H)(H)\cdots Br]^{-} \longrightarrow HO—CH_2CH_3 + :Br^{-}$$

Transition state

With a tertiary halogenoalkane, the mechanism is still nucleophilic substitution:

$$(CH_3)_3CBr + OH^- \rightarrow (CH_3)_3COH + Br^-$$

However, in this mechanism, only *one* molecule is present in the rate-determining or slow step. This is an S_N1 mechanism, although there are still two stages:

$$(CH_3)_3C—Br \xrightleftharpoons{\text{Slow}} (CH_3)_3C^+ + :Br^-$$

$$(CH_3)_3C^+ + :\bar{O}H \xrightarrow{\text{Fast}} (CH_3)_3C—OH$$

Tip Think of S_N1 as 'Substitution Nucleophilic, 1 reactant'.

For a secondary halogenoalkane, the mechanism is a combination of S_N1 and S_N2.

There is another reaction that can take place when halogenoalkanes react with hydroxide ions. This is covered on page 142.

Formation of nitriles

When a halogenoalkane is heated under reflux with cyanide ions dissolved in ethanol, the cyanide ion is substituted for the halogen and a nitrile is formed:

$$CH_3CH_2Br + CN^- \rightarrow CH_3CH_2CN + Br^-$$

This is an important reaction because a carbon atom has been added to the carbon chain and the nitrile group can be reacted further.

Secondary and tertiary halogenoalkanes react in a similar way, but the mechanism may be different.

Formation of primary amines

When a halogenoalkane is heated with ammonia, a reflux process cannot be used because the ammonia would escape as gas. This reaction has to be carried out in a sealed tube. The reaction takes place in two steps:

$$CH_3CH_2Br + NH_3 \rightarrow CH_3CH_2NH_3^+Br^-$$

$$CH_3CH_2NH_3^+Br^- + NH_3 \rightarrow CH_3CH_2NH_2 + NH_4^+Br^-$$

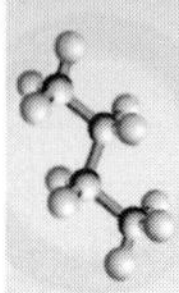

However, in a sealed tube the reaction does not stop with the formation of a primary amine. It continues replacing successive hydrogens on the nitrogen to give secondary and tertiary amines, and finally a quaternary ammonium salt. Note that these reactions are *not* needed for A-level.

Elimination of hydrogen bromide

Under similar conditions to those needed for nucleophilic substitution, it is possible to get an elimination reaction to take place. If, instead of aqueous ethanol, a concentrated hydroxide solution in pure ethanol is used, HBr is eliminated and a double bond is formed:

$$CH_3CH_2Br + OH^- \rightarrow CH{=}CH_2 + H_2O + Br^-$$

It is important to remember that different halogenoalkanes favour one type of reaction over the other (Table 15.1).

Table 15.1

Halogenoalkane	Reaction favoured
Primary	Mainly substitution
Secondary	Both
Tertiary	Mainly elimination

For a given halogenoalkane, to favour **substitution** use:
- lower temperatures
- more dilute solutions of sodium or potassium hydroxide
- more water in the solvent mixture

To favour **elimination** use:
- higher temperatures
- a concentrated solution of sodium or potassium hydroxide
- pure ethanol as the solvent

Different types of halogenoalkane

So far we have used bromoethane as the example of a halogenoalkane and for most purposes that is fine. However, we must not forget that the different halogens have an effect on the halogenoalkanes they form, not least because of the relative strengths of the C—Hal bond (Table 15.2).

Table 15.2

Bond	Bond energy/$kJ\,mol^{-1}$
C—F	467
C—Cl	338
C—Br	276
C—I	238

For a halogenoalkane to react, the C—Hal bond has to be broken. From Table 15.2 you can see that this is much more difficult for fluoroalkanes than for the other members of the group.

A number of halogenoalkanes have important uses and in looking at these we need to bear this factor in mind.

Uses

The chemical inertness of chlorofluorocarbons (CFCs) has meant that until recently they were used as propellants in aerosols, as refrigerant gases in refrigerators, as 'blowing agents' for making plastic foams (such as expanded polystyrene) and as solvents. There has been considerable recent evidence of their harmful effects on the atmosphere, which has resulted in a large decrease in their use.

CFCs are largely responsible for destroying the ozone layer. In the high atmosphere, ultraviolet light causes carbon-to-chlorine bonds to break forming chlorine free radicals. It is these free radicals that destroy ozone.

It has also been shown that CFCs can cause global warming. One molecule of CCl_3F, for example, has a global warming potential about 5000 times greater than a molecule of carbon dioxide. Fortunately, there is much less of this compound than carbon dioxide in the atmosphere.

The halogen-containing plastics PVC (poly(chloroethene)) and PTFE (poly(tetrafluoroethene)) are commercially important. The plastics themselves are not halogenoalkanes, but they are made from halogenoalkenes.

As well as these major uses, halogenoalkanes are used as flame retardants and as anaesthetics. These uses once again rely on the relative inertness of halogenoalkanes.

Chlorobenzene is a halogenoarene:

It is much less reactive towards nucleophilic substitution than the halogenoalkanes. The C—Cl bond in the molecule is stronger than you might expect. This is because one of the lone pairs of electrons on the chlorine atom is delocalised with the ring electrons on benzene:

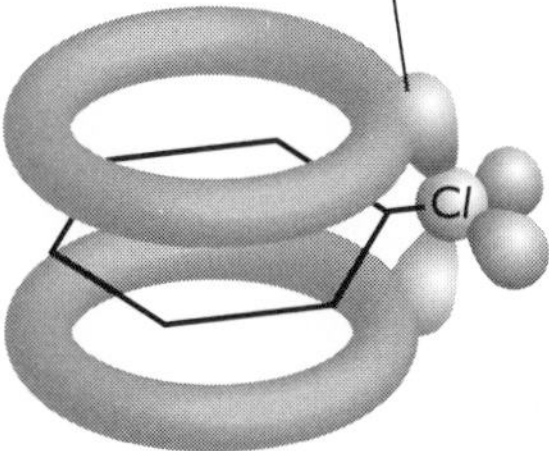

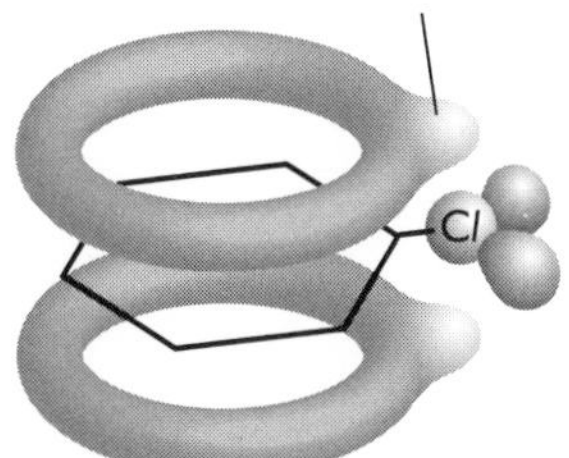

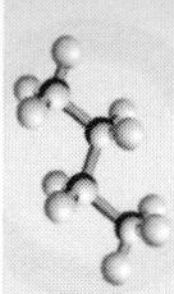

Hydroxy compounds

The first part of this section dealing with alcohols is needed for the AS examination. The second part, dealing with the iodoform reaction, phenol and the relative acidities of water, phenol and ethanol is needed for the A2 examination.

Different types of alcohol

In just the same way as with primary, secondary and tertiary halogenoalkanes, there are primary, secondary and tertiary alcohols.

Ethanol, CH_3CH_2OH, is a primary alcohol.

Propan-2-ol, $CH_3CHOHCH_3$, is a secondary alcohol.

2-methylpropan-2-ol is a tertiary alcohol:

```
        CH3
        |
  CH3—C—OH
        |
        CH3
```

Reactions of alcohols

Combustion

Like most organic compounds, alcohols are flammable. You may remember this from using a spirit burner. The equation for the complete combustion of ethanol is:

$$CH_3CH_2OH + 3O_2 \rightarrow 2CO_2 + 3H_2O$$

Substitution to form halogenoalkanes

Halogenoalkanes can be hydrolysed to give alcohols, and alcohols can be converted into halogenoalkanes, but by different reagents.

One way of carrying out the substitution is to use the appropriate hydrogen halide. This method works for the bromo- and iodoalkanes if the hydrogen halide is generated in the reaction flask. Sodium bromide with concentrated sulfuric acid can be used for the bromoalkane, but sodium iodide and concentrated phosphoric(V) acid have to be used for the iodoalkane because sulfuric acid would oxidise any HI formed. The equations for the formation of bromoethane and iodoethane are as follows:

$$CH_3CH_2OH + HBr \rightarrow CH_3CH_2Br + H_2O$$

$$CH_3CH_2OH + HI \rightarrow CH_3CH_2I + H_2O$$

The following method only works with tertiary alcohols, forming the tertiary chloroalkane:

$$\begin{array}{c}CH_3\\|\\CH_3{-}C{-}OH\\|\\CH_3\end{array} + HCl \longrightarrow \begin{array}{c}CH_3\\|\\CH_3{-}C{-}Cl\\|\\CH_3\end{array} + H_2O$$

For other chloroalkanes we have to use phosphorus trichloride, PCl_3, phosphorus pentachloride, PCl_5, or thionyl chloride, $SOCl_2$:

$$3CH_3CH_2OH + PCl_3 \rightarrow 3CH_3CH_2Cl + H_3PO_3$$

$$CH_3CH_2OH + PCl_5 \rightarrow CH_3CH_2Cl + POCl_3 + HCl$$

$$CH_3CH_2OH + SOCl_2 \rightarrow CH_3CH_2Cl + SO_2 + HCl$$

Reaction with sodium

When a small piece of sodium is dropped into ethanol it dissolves, producing bubbles of hydrogen gas. It leaves a colourless solution which, if evaporated to dryness, produces a white solid. This white solid is sodium ethoxide, $CH_3CH_2O^-Na^+$:

$$2CH_3CH_2OH + 2Na \rightarrow 2CH_3CH_2O^-Na^+ + H_2$$

This reaction is sometimes used to dispose of small amounts of old sodium because it is much less violent than reacting sodium with water. It can also be used as a test for an alcohol. The ethoxide ion (like the hydroxide ion) is a strong base and a good nucleophile.

Oxidation reactions

In the presence of an oxidising agent such as acidified dichromate(VI) solution, whether or not an alcohol is oxidised depends on its structure. A positive test for oxidation is that the dichromate(VI) solution turns from orange to blue-green.

- On warming a primary alcohol with acidified dichromate(VI), an aldehyde is first formed. If this is not removed from the reaction vessel it is further oxidised to a carboxylic acid. The mixture turns from orange to blue-green:

$$\begin{array}{c}H\quad [O]\\|\\CH_3{-}C{-}O{-}H\\|\\H\end{array} \longrightarrow CH_3{-}C\begin{array}{l}{=}O\\{-}H\end{array} + H_2O$$

$$CH_3{-}C\begin{array}{l}{=}O\\{-}H\end{array} + [O] \longrightarrow CH_3{-}C\begin{array}{l}{=}O\\{-}OH\end{array}$$

- On refluxing a secondary alcohol with acidified dichromate(VI), a ketone is formed, which is not oxidised further. Again the mixture turns from orange to blue-green:

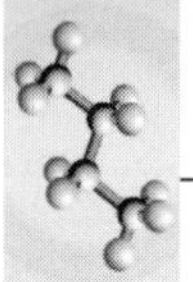

$$CH_3-\underset{\underset{CH_3}{|}}{\overset{\overset{H}{|}}{C}}-O-H \xrightarrow{[O]} CH_3-\underset{CH_3}{C}=O + H_2O$$

Propan-2-ol Propanone

- With a tertiary alcohol, there are no hydrogen atoms on the carbon atoms that can be oxidised, so there is no reaction.

Tertiary alcohol

$$CH_3-\underset{\underset{CH_3}{|}}{\overset{\overset{CH_3}{|}}{C}}-O-H \quad [O]$$

There is no hydrogen attached to this carbon for the oxygen to remove.

Provided that an aldehyde and a ketone can be distinguished between, these oxidation reactions can be used to detect primary, secondary or tertiary alcohols. There are relatively simple tests for aldehydes that use Fehling's solution or Tollens' reagent (see Chapter 16).

Dehydration

Strong acids such as phosphoric(V) and sulfuric can be used to dehydrate alcohols to form alkenes:

$$CH_3CH_2CH_2OH \rightarrow CH_3CH{=}CH_2 + H_2O$$

Warming ethanol and passing the vapour over heated aluminium oxide achieves the same reaction:

$$CH_3CH_2OH \rightarrow CH_2{=}CH_2 + H_2O$$

Forming esters

If an alcohol is warmed with an organic acid in the presence of H^+ ions, an ester is formed with the elimination of water:

$$CH_3CH_2-C(=O)-O-H + CH_3CH_2OH \rightleftharpoons CH_3-C(=O)-O-CH_2CH_3 + H_2O$$

This is an equilibrium reaction and it is often quite slow.

The reaction can be speeded up by using another compound containing the RC=O group. Suitable compounds include acyl chlorides or acid anhydrides (see Chapter 17).

With ethanoyl chloride and ethanol, the following reaction occurs:

$$CH_3COCl + CH_3CH_2OH \rightarrow CH_3CO_2CH_2CH_3 + HCl$$

With ethanoic anhydride and ethanol, this reaction takes place:

$$(CH_3CO)_2O + CH_3CH_2OH \rightarrow CH_3CO_2CH_2CH_3 + CH_3CO_2H$$

In naming esters the convention is to give the fragment of the alcohol first and then the anion name of the acid. So the esters shown above are all ethyl ethanoate (rather strangely they are usually drawn with the acid fragment first!).

Try this yourself

(36) Name the four esters whose structures are shown below:

(a) $CH_3CH_2—C(=O)—O—CH_2CH_3$

(b) $H—C(=O)—O—CH_2CH_2CH_3$

(c) $CH_3CH_2CH_2—C(=O)—O—CH_3$

(d) $CH_3—C(=O)—O—CH_2CH_2CH_3$

The tri-iodomethane (iodoform) reaction

This reaction is quite specific for a particular structural arrangement in an alcohol. It detects the presence of the $CH_3CH(OH)–$ group. The test is carried out by adding iodine solution to the alcohol and then adding just enough sodium hydroxide to remove the colour of the iodine. On standing, or more usually on warming, a pale yellow precipitate of tri-iodomethane is formed if the group is present.

Phenol

Phenol is an aromatic hydroxyl compound:

OH

The presence of the benzene ring makes phenol behave differently from alcohols.

Reactions with bases

Phenol is a weakly acidic compound. Its acidity stems from the fact that it ionises in water:

OH + H_2O ⇌ O^- + H_3O^+

Phenoxide ion

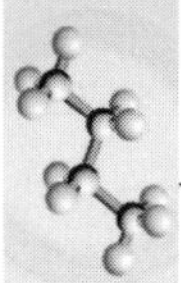

Because phenol is a very weak acid, the equilibrium lies over to the left. It can lose a proton because the remaining negative charge is delocalised over the benzene ring, making the phenoxide ion more stable. There is evidence for this behaviour because phenol reacts with sodium hydroxide to give a colourless product:

$$C_6H_5OH + NaOH \longrightarrow C_6H_5O^-Na^+ + H_2O$$

Sodium phenoxide

But phenol is neither acidic enough to turn blue litmus paper red, nor to release carbon dioxide from sodium carbonate. It does produce hydrogen when heated with sodium:

$$2C_6H_5OH + 2Na \longrightarrow 2C_6H_5O^-Na^+ + H_2$$

Sodium phenoxide

Nitration

Phenol behaves differently from benzene in its reaction with nitric acid. It reacts in the cold with dilute nitric acid, whereas benzene requires a nitrating mixture of concentrated nitric and sulfuric acids. The reason for this is that the presence of the OH group makes the ring much more reactive. It also directs reaction to the 2- and 4-positions on the ring.

With dilute nitric acid, 4-nitrophenol is formed:

$$C_6H_5OH + HNO_3 \xrightarrow[\text{Room temp.}]{\text{Dilute acid}} HOC_6H_4NO_2 + H_2O$$

4-nitrophenol

With concentrated nitric acid, 2,4,6-trinitrophenol is formed:

$$C_6H_5OH + 3HNO_3 \xrightarrow{\text{Conc. acid}} HOC_6H_2(NO_2)_3 + 3H_2O$$

2,4,6-trinitrophenol

Bromination

When bromine water is added to phenol, there is a similar effect to nitration. The activated ring gives an almost instantaneous white precipitate of 2,4,6-tribromophenol:

$$C_6H_5OH + 3Br_2 \xrightarrow{\text{Room temp.}} C_6H_2Br_3OH + 3HBr$$

2,4,6-tribromophenol

16 Carbonyl compounds

This chapter is needed almost exclusively for the AS examination. Only the very last section is required for the A2 examination. The syllabus treats aldehydes and ketones as two separate classes of organic compound. However, the learning outcomes and this section deal with them together because many of the reactions are similar.

Formation

The oxidation of alcohols by acidified dichromate(VI) was looked at in Chapter 15.

- With a primary alcohol, an aldehyde is first formed. If this is not removed from the reaction vessel, it is further oxidised to give a carboxylic acid:

$$CH_3CH_2OH + [O] \longrightarrow CH_3CHO + H_2O$$

$$CH_3CHO + [O] \longrightarrow CH_3COOH$$

- With a secondary alcohol, a ketone is formed. This is not oxidised further:

$$CH_3CH(OH)CH_3 + [O] \longrightarrow CH_3COCH_3 + H_2O$$

Propan-2-ol → Propanone

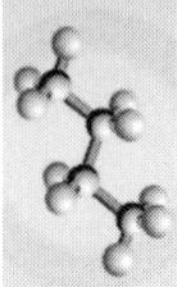

Reactions

Reduction

This is the reverse of the oxidation reactions used in the preparation of carbonyl compounds. It is carried out using sodium tetrahydridoborate (sodium borohydride), $NaBH_4$.

With aldehydes such as ethanal the reaction is as follows, forming a primary alcohol:

$$CH_3{-}C(=O){-}H + 2[H] \longrightarrow CH_3{-}C(OH)(H){-}H \quad (CH_3CH_2OH)$$

With ketones such as propanone the reaction is very similar. A secondary alcohol is formed:

$$(CH_3)_2C{=}O + 2[H] \longrightarrow CH_3{-}C(OH)(CH_3){-}H \quad (CH_3CH(OH)CH_3)$$

Nucleophilic addition of hydrogen cyanide

In just the same way that alkenes can react by adding a molecule across a C=C double bond, carbonyl compounds can add a molecule across the C=O double bond. At AS you need to know the mechanism for this reaction, using hydrogen cyanide as the nucleophile. You may think that this is a strange reactant, but as we saw earlier it adds a carbon atom to the chain, which is often an important step in a reaction.

The reactions can be summarised in simple equations:

$$CH_3{-}C(=O){-}H + HCN \longrightarrow CH_3{-}C(OH)(H){-}CN$$

$$(CH_3)_2C{=}O + HCN \longrightarrow CH_3{-}C(OH)(CH_3){-}CN$$

Hydrogen cyanide itself is not used in the reaction because it is a highly toxic gas. Instead, sodium or potassium cyanide is added to the carbonyl compound followed by a small amount of sulfuric acid. This produces hydrogen cyanide in the reaction vessel but also forms cyanide ions, which are important as you will see when we look at the mechanism.

The first thing to remember is that the C=O bond is polarised with a partial positive charge on the carbon and a partial negative charge on the oxygen:

$$\mathrm{C^{\delta+}{=}O^{\delta-}}$$

Mechanism of addition of hydrogen cyanide to an aldehyde

The reaction starts with an attack by the nucleophilic cyanide ion on the slightly positive carbon atom:

$$\mathrm{CH_3{-}C^{\delta+}H{=}O^{\delta-} + HCN \longrightarrow CH_3{-}C(O^-)(H)(CN)}$$

The negative ion formed then picks up a hydrogen ion. It could come from a hydrogen cyanide molecule or from the water or the H_3O^+ ions present in the slightly acidic solution:

$$\mathrm{CH_3{-}C(O^-)(H)(CN) + H{-}CN \longrightarrow CH_3{-}C(OH)(H)(CN) + {:}CN^-}$$

Mechanism of addition of hydrogen cyanide to a ketone

The mechanism for ketones is similar to that for aldehydes. The first stage is a nucleophilic attack by the cyanide ion on the slightly positive carbon atom:

$$\mathrm{(CH_3)_2C^{\delta+}{=}O^{\delta-} + 2[H] \longrightarrow CH_3{-}C(CH_3)(O^-)(CN)}$$

The negative ion formed then picks up a hydrogen ion to give the hydroxynitrile (cyanohydrin as it is sometimes called in older textbooks).

$$\mathrm{CH_3{-}C(CH_3)(O^-)(CN) + H{-}CN \longrightarrow CH_3{-}C(CH_3)(CN){-}O{-}H + {:}CN^-}$$

Chirality

You may have noticed that the product formed with the aldhyde has four different groups attached to the central carbon. If you remember the section on isomerism in Chapter 13 you will know that this is therefore a **chiral** carbon atom.

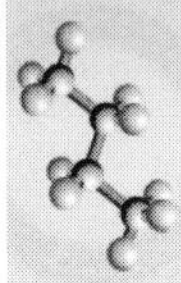

You may also remember that this usually means that the compound exists as a pair of optical isomers that are mirror images of each other. The product of this particular reaction, however, is *not* optically active because ethanal is a planar molecule and the mechanism means that attack by the cyanide ion can occur from both above and below the molecule. This produces a 50:50 mixture of isomers, so the net result is a lack of optical activity.

Reaction with 2,4-dinitrophenylhydrazine (2,4-DNPH)

This rather complicated sounding reagent is used as a simple test for carbonyl compounds. The reaction is called a 'condensation' or sometimes a 'nucleophilic addition–elimination' reaction.

If a few drops of a suspected carbonyl compound (or a solution of the suspected carbonyl compound in methanol) are added to 2,4-DNPH, a distinct orange or yellow precipitate shows a positive result.

$$R_2C{=}O + H_2N{-}NH{-}C_6H_3(NO_2)_2 \longrightarrow R_2C{=}N{-}NH{-}C_6H_3(NO_2)_2 + H_2O$$

The equation for the reaction is shown above, but it is unlikely that you would be asked to produce this in the examination.

The reaction is rather more useful than just testing for a carbonyl compound. If the precipitate is filtered off, washed and re-crystallised, the melting point of the crystals obtained is characteristic of the particular aldehyde or ketone that reacted, enabling identification to take place.

Distinguishing between aldehydes and ketones

In the section on oxidation reactions of hydroxy compounds in Chapter 15, you saw that primary, secondary and tertiary alcohols can be distinguished between by looking at their oxidation products. These same reactions enable us to distinguish between aldehydes and ketones.

Table 16.1 summarises the behaviour of the two types of carbonyl compound with different reagents.

Table 16.1

Reagent	Aldehydes	Ketones
Acidified dichromate(VI)	Orange solution turns blue-green	No change in the orange solution
Fehling's solution	Blue solution produces an orange-red precipitate of copper(I) oxide	No change in the blue solution
Tollens' reagent	Colourless solution produces a grey precipitate of silver, or a silver mirror is formed on the test tube	No change in the colourless solution

Try this yourself

(37) Compound **K** was produced by the oxidation of an alcohol **J** of molecular formula $C_4H_{10}O$. When **K** is reacted with 2,4-DNPH, a yellow-orange precipitate is formed. **K** also reacts with Fehling's solution forming an orange-red precipitate. When treated with alkaline aqueous iodine, no precipitate is formed.

Study the reactions above and use them to deduce the structural formulae of **J** and **K**.

The tri-iodomethane (iodoform) reaction

We saw in Chapter 15 that this reaction is linked to the $CH_3CH(OH)-$ group in alcohols. It can also be used to identify the CH_3CO- group in carbonyl compounds.

In other words, a positive result — the pale yellow precipitate of tri-iodomethane (iodoform) — is given by an aldehyde or ketone containing the group:

$$CH_3-C(=O)-R$$

Ethanal is the *only* aldehyde to give a positive reaction. Any methyl ketone will give a positive result.

$$CH_3-C(=O)-R + 3I_2 + 3OH^- \longrightarrow CI_3-C(=O)-R + 3I^- + 3H_2O$$

$$CI_3-C(=O)-R + OH^- \longrightarrow CHI_3 + R-C(=O)-O^-$$

This bond is broken

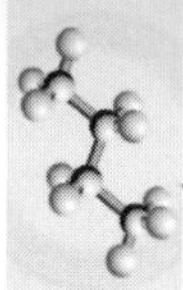

17 Carboxylic acids and their derivatives

This chapter is a mix of material for the AS and A2 examinations. Most of the material on carboxylic acids and esters is for the AS examination; the material related to other derivatives is for A2.

Carboxylic acids

Formation of carboxylic acids

There are three main methods for preparing carboxylic acids (although, strictly, one is a step in another process).

From an alcohol

You will remember from Chapter 15 that a primary alcohol can be oxidised using acidified dichromate(VI) to give an aldehyde and that if this is not distilled off, it is oxidised further to a carboxylic acid, for example:

$$CH_3-CH_2-O-H \xrightarrow{[O]} CH_3-CHO + H_2O$$

From an aldehyde

We could start with an aldehyde and oxidise it with acidified dichromate(VI) to form the carboxylic acid, for example:

$$CH_3-CHO + [O] \longrightarrow CH_3-COOH$$

This is rare, since in order to make an aldehyde we have to start with a primary alcohol.

From a nitrile

You saw in Chapter 15 how nitriles can be produced from halogenoalkanes. They can be hydrolysed to form either carboxylic acids or their salts, depending on whether acid or base is used in the hydrolysis.

If the nitrile is heated under reflux with a dilute acid, such as dilute hydrochloric acid, a carboxylic acid is formed, for example:

$$CH_3CN + 2H_2O + H^+ \rightarrow CH_3CO_2H + NH_4^+$$

If the nitrile is heated under reflux with an alkali such as sodium hydroxide solution, the sodium salt of the carboxylic acid is formed and ammonia is released, for example:

$$CH_3CN + NaOH + H_2O \rightarrow CH_3CO_2^-Na^+ + NH_3$$

To obtain the carboxylic acid, a strong acid such as hydrochloric acid is added and the carboxylic acid is distilled off.

Reactions of carboxylic acids

Formation of salts

Carboxylic acids are generally relatively weak acids (although there are exceptions). They behave as acids because of their ability to donate protons:

$$CH_3CO_2H + H_2O \rightleftharpoons CH_3CO_2^- + H_3O^+$$

The reaction is reversible with the equilibrium well over to the left. Ethanoic acid is never more than around 1% ionised resulting in solutions with a pH of between 2 and 3.

As a result, carboxylic acids can form salts in a number of ways (although there are some exceptions). Ethanoic acid is used as the example in all the equations shown.

- Aqueous solutions of carboxylic acids react with the more reactive metals such as magnesium to form the salt:

$$2CH_3CO_2H + Mg \rightarrow (CH_3CO_2^-)_2Mg^{2+} + H_2$$

- Aqueous solutions of carboxylic acids react with metal hydroxides such as sodium hydroxide to form the salt:

$$CH_3CO_2H + NaOH \rightarrow CH_3CO_2^-Na^+ + H_2O$$

- Aqueous solutions of carboxylic acids react with carbonates and hydrogen carbonates liberating carbon dioxide:

$$2CH_3CO_2H + Na_2CO_3 \rightarrow 2CH_3CO_2^-Na^+ + CO_2 + H_2O$$

$$CH_3CO_2H + NaHCO_3 \rightarrow CH_3CO_2^-Na^+ + CO_2 + H_2O$$

  There is very little difference between these reactions and those with other acids. If, however, you chose to use a marble chip as the carbonate, the reaction would be noticeably slower.

- Ethanoic acid reacts with ammonia in just the same way as other acids, forming the ammonium salt:

$$CH_3CO_2H + NH_3 \rightarrow CH_3CO_2^-NH_4^+$$

Formation of esters

Making esters from alcohols was covered in Chapter 15, and you learned that a range of reagents can be used to react with the –OH group. Here, we are considering a single reaction — the formation of an ester from a carboxylic acid and an alcohol.

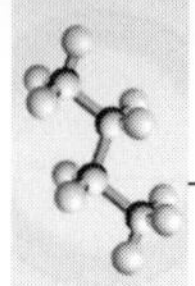

An ester is formed when a carboxylic acid is heated with an alcohol in the presence of an acid catalyst, usually concentrated sulfuric acid. The reaction is slow and reversible, for example:

$$CH_3CH_2-C(=O)-O-H + CH_3CH_2OH \rightleftharpoons CH_3-C(=O)-O-CH_2CH_3 + H_2O$$

In the laboratory this is achieved by warming the carboxylic acid and alcohol together with a few drops of concentrated sulfuric acid in a water bath for 10–15 minutes and then pouring the contents into a small beaker of cold water. The ester can be detected by its fruity smell. If a sample is required, it can be distilled off from the reaction mixture. Ethyl ethanoate, the most common ester you will come across, can be prepared in this way.

Formation of acyl chlorides

To form an acyl halide from a carboxylic acid, the –OH group in the acid has to be replaced by a –Hal group. You might wonder why this is an important reaction, but acyl chlorides are very reactive and are useful in preparing a range of new materials.

In Chapter 15 we used a group of reagents — phosphorus trichloride, PCl_3, phosphorus pentachloride, PCl_5, and thionyl chloride, $SOCl_2$ — to convert the –OH group of an alcohol to a –*Cl* group. This same group of reagents can be used to achieve the same outcome here:

$$3CH_3CO_2H + PCl_3 \rightarrow 3CH_3COCl + H_3PO_3$$

$$CH_3CO_2H + PCl_5 \rightarrow CH_3COCl + POCl_3 + HCl$$

$$CH_3CO_2H + SOCl_2 \rightarrow CH_3COCl + SO_2 + HCl$$

The third reaction is the 'cleanest' because the by-products are gases.

Relative acidity of acids

For the A2 examination it is not enough just to know that carboxylic acids are relatively weak acids because the equilibrium for dissociation lies well to the left. You also need to know what makes some acids stronger or weaker than others.

It helps if you think back to the definition of an acid as a proton donor given in Chapter 7.

Tip It would be a good idea to re-read the section on pK_a and pH on page 73.

Before comparing carboxylic acids, it is useful to look at the ionisation of ethanoic acid in more detail:

$$CH_3-C(=O)-O-H \longrightarrow CH_3-C(=O)-O^- + H^+$$

It is helpful to look a little more closely at the ethanoate ion. It has been found that the two carbon–oxygen bond lengths are the same. This means that the usual way of drawing the ion with a C═O and a C—O⁻ cannot be correct. The representation below is more accurate:

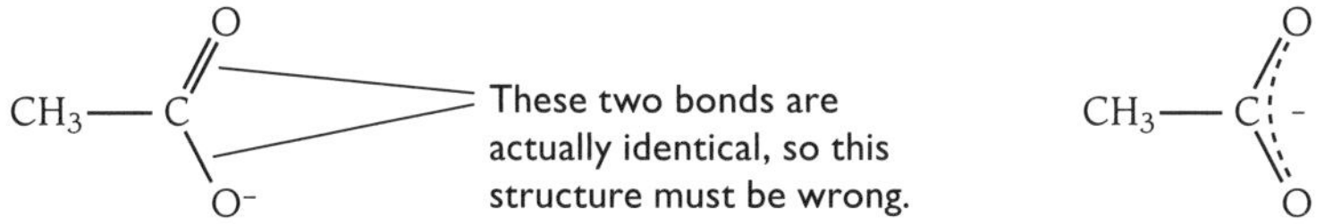

CH_3—C(═O)—O^- These two bonds are actually identical, so this structure must be wrong.

CH_3—C(…O)(…O) -

The dashed line represents delocalisation of electrons over the two oxygen atoms and the carbon atom. In general, the more the charge is spread around, the more stable the ion is.

The pK_a values for some carboxylic acids are given in Table 17.1.

Table 17.1

Formula of the acid	**pK_a**
HCO_2H	3.75
CH_3CO_2H	4.76
$CH_3CH_2CO_2H$	4.87

You might be surprised to see that methanoic acid has a smaller pK_a (stronger acid) than ethanoic acid. Let's think what we know about alkyl groups. They have a tendency to release electrons, which reduces the overall negative charge on the O—C—O group. However, methanoic acid does not have any alkyl groups, so this effect is not present. The addition of an extra CH_2 in propanoic acid makes very little difference to the pK_a.

Now let's see what happens if an **electronegative** atom that pulls electrons away from the O—C—O group is introduced. Chlorine is a good example of an electronegative atom. The pK_a values for ethanoic acid and its chlorinated derivatives are shown in Table 17.2.

Table 17.2

Formula of the acid	**pK_a**
CH_3CO_2H	4.76
CH_2ClCO_2H	2.86
$CHCl_2CO_2H$	1.29
CCl_3CO_2H	0.65

You can see how the presence of chlorine atoms makes it easier for the hydrogen to be removed.

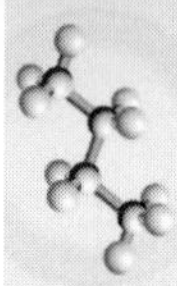

Acyl chlorides

We saw earlier how to make acyl chlorides from carboxylic acids. Here, their reactions are covered.

Hydrolysis

Acyl halides react quite violently with water, releasing steamy fumes of HCl and forming the appropriate carboxylic acid:

$$CH_3-C(=O)-Cl \; + \; H-O-H \longrightarrow CH_3-C(=O)-O-H \; + \; HCl$$

Figure 17.1 The reaction of ethanoyl chloride with water

The equation can also be written as shown below:

$$CH_3COCl + H_2O \rightarrow CH_3CO_2H + HCl$$

However, when studying the reactions of acyl chlorides with alcohols and phenols, it is useful to think about the form shown in Figure 17.1.

The ease with which acyl chlorides react with water is in contrast with the reactivity of alkyl and aryl chlorides. In Chapter 15 you saw that alkyl halides can be hydrolysed by heating under reflux with sodium hydroxide in aqueous ethanol.

Aryl chlorides are very resistant to hydrolysis. Chlorobenzene shows signs of reaction with OH^- only under extreme conditions of around 500 K and approaching 200 atm pressure. One reason for this lack of reaction is repulsion of the OH^- group by the ring electrons. Perhaps a more significant reason is the interaction between one of the lone pairs of electrons on chlorine and the delocalised ring electrons:

Overlap between lone pair and the ring electrons

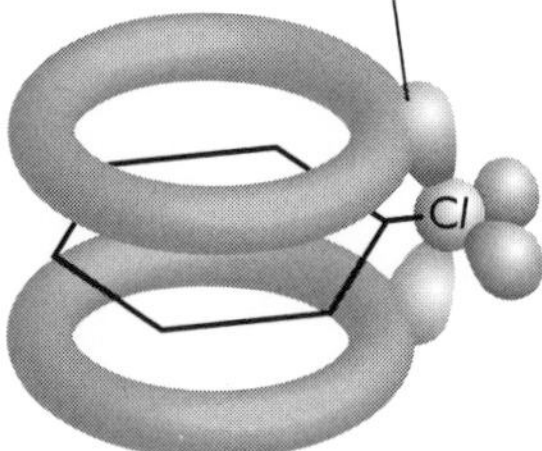

Lone pair now delocalised to some extent with the ring electrons

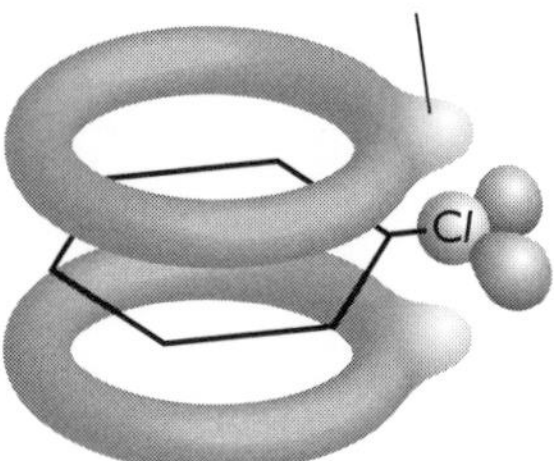

Reaction with alcohols and phenols

With alcohols

The reaction between an acyl chloride and an alcohol is similar to the reaction between an acyl chloride and water, with the release of hydrogen chloride:

$$CH_3-C(=O)-Cl \; + \; H-O-CH_2CH_3 \longrightarrow CH_3-C(=O)-O-CH_2CH_3 \; + \; HCl$$

Compare this reaction with the equation shown in Figure 17.1.

From a practical point of view this is a straightforward method for producing an ester because it takes place at room temperature and, because HC*l* is released, it is not a reversible reaction. This contrasts with the reaction of ethanol with ethanoic acid in the presence of an acid catalyst.

With phenols

In a phenol, the –OH group is attached directly to the benzene ring. The reaction between phenol and ethanoyl chloride is not as vigorous as the reaction with ethanol due to the presence of the benzene ring. Apart from this the reaction is very similar:

$$CH_3-C(=O)-Cl \; + \; H-O-C_6H_5 \longrightarrow CH_3-C(=O)-O-C_6H_5 \; + \; HCl$$

The product is called phenyl ethanoate (remember alcohol/phenol part first, then the carboxylic acid part). The reaction is called **acylation**.

Reaction with primary amines

A primary amine is an ammonia molecule in which one of the hydrogen atoms has been replaced by an alkyl group:

NH_3	CH_3NH_2	$CH_3CH_2NH_2$	$C_6H_5NH_2$
Ammonia	Methylamine	Ethylamine	Phenylamine

Primary amines react with acyl chlorides by substitution, replacing the chlorine atom and joining via the nitrogen atom, for example:

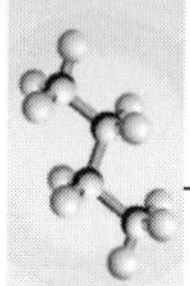

$$CH_3C(=O)Cl + H{-}N(CH_2CH_3){-}H \longrightarrow CH_3C(=O){-}N(H){-}CH_3 + HCl$$

The product is an **N-substituted amide**.

Esters

Most of this section is needed for AS; only polyesters are required for A2.

Hydrolysis

You will remember that to form an ester, an alcohol (or phenol) is reacted with a carboxylic acid and water is formed as a by-product. Hydrolysis is the reverse of this process and can be achieved in either of two ways.

Acid hydrolysis

In acid hydrolysis, the ester is heated under reflux with either dilute hydrochloric or dilute sulfuric acid:

$$CH_3CO_2CH_2CH_3 + H_2O \rightleftharpoons CH_3CO_2H + CH_3CH_2OH$$

As in the formation of an ester, the reaction is reversible. We try to make the reaction as complete as possible by having an excess of water present.

Base hydrolysis

Base hydrolysis is the more usual way of hydrolysing esters because the reaction goes to completion, rather than forming an equilibrium mixture:

$$CH_3CO_2CH_2CH_3 + NaOH \rightarrow CH_3CO_2^-Na^+ + CH_3CH_2OH$$

Although this forms the sodium salt of the carboxylic acid, it is still relatively easy to separate the products. First, the alcohol is distilled off and then an excess of a strong acid (dilute hydrochloric or dilute sulfuric acid) is added. This forms the carboxylic acid which is only slightly ionised and can be distilled off.

Commercial uses

Perfumes and flavours

Esters can be formed from a large number of combinations of alcohols with carboxylic acids. They are distributed widely in nature and are responsible for many of the smells and flavours associated with fruits and flowers. This has led to the development of artificial esters for the food industry — for example for ice cream — and for products that need a nice smell, such as detergents and air fresheners. Figure 17.2 shows some of the esters that are used.

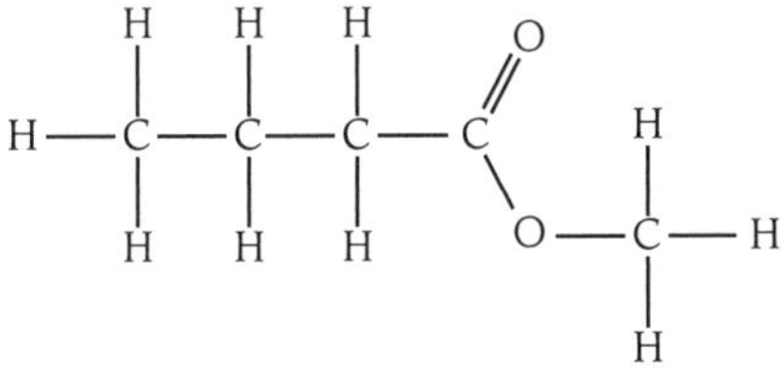

Methyl butanoate (apple)

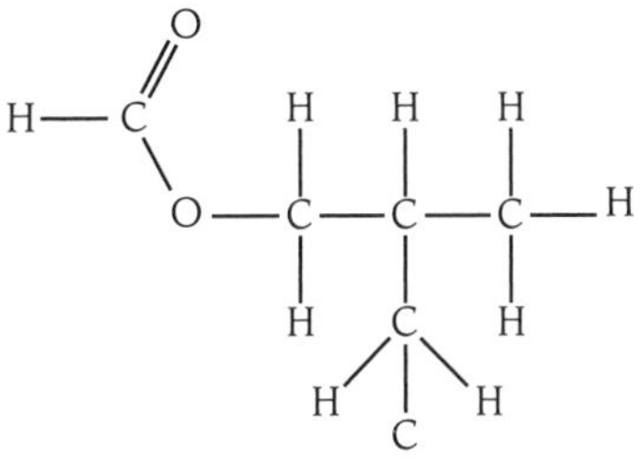

Methylpropyl methanoate (raspberry)

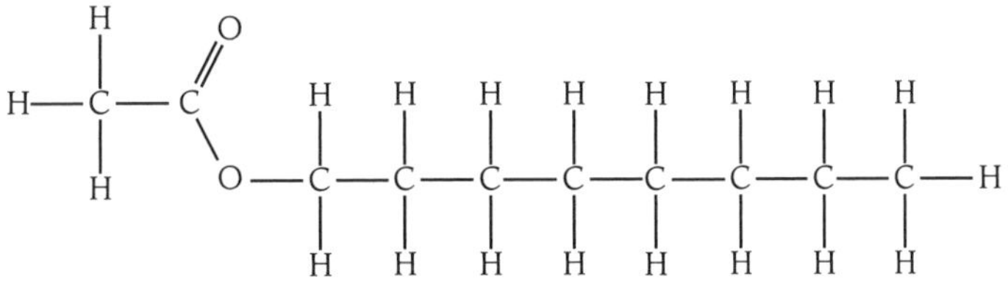

Octyl ethanoate (orange)

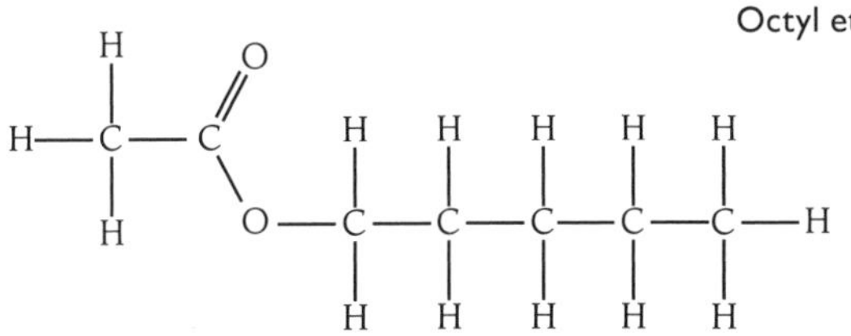

Pentyl ethanoate (pear)

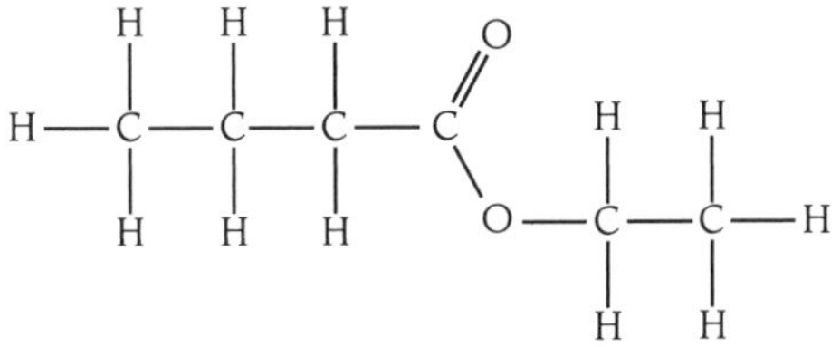

Ethyl butanoate (pineapple)

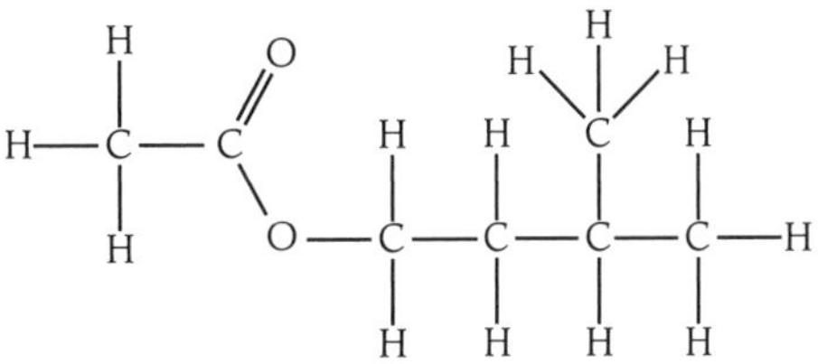

3-methylbutyl ethanoate (banana)

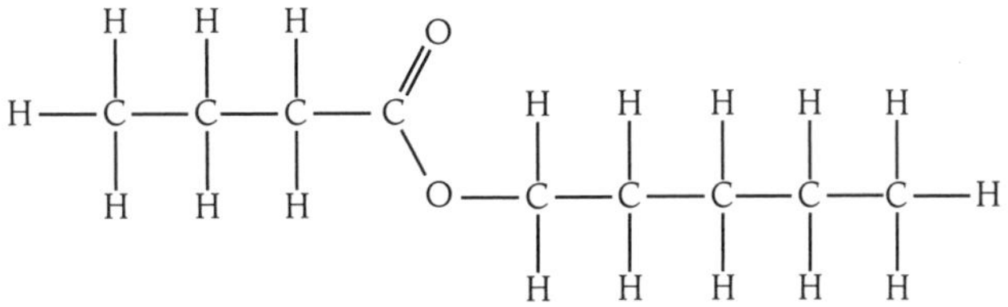

Pentyl butanoate (apricot, strawberry)

Figure 17.2 Esters used in industry

Natural esters are still used in perfumes.

Margarine production

Most oils and fats are esters of long-chain carboxylic acids with the alcohol glycerol ($CH_2OHCHOHCH_2OH$). Oils often contain carboxylic acids in which the chain has one or more C═C double bonds present. Fats contain mainly carboxylic acids with saturated chains (containing only C—C single bonds) (see Figure 17.3).

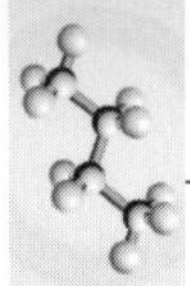

It is this that makes the oils liquid and the fats low-melting solids. These esters are often referred to as glycerides because the common name for the alcohol is glycerol.

$$\begin{array}{r} CH_3(CH_2)_{16}COOCH_2 \\ | \\ CH_3(CH_2)_{16}COOCH \\ | \\ CH_3(CH_2)_{16}COOCH_2 \end{array}$$

Figure 17.3 A molecule of a fat

In order to make margarine, the vegetable oils have to have some of the double bonds removed. This is achieved by reacting the oils with hydrogen in the presence of a nickel catalyst.

Consumption of saturated fats has been linked to heart disease, so it is important that these do not form a large part of the human diet. Saturated fats are more common in animal fat than in vegetable oils.

Soap making

Animal fats and vegetable oils have been the raw materials for making soaps and some detergents for hundreds of years. The large molecule esters found in oils and fats are heated with concentrated sodium hydroxide solution and undergo base hydrolysis. The sodium salt of the carboxylic acid is formed (and acts as the soap or detergent) along with the alcohol 1,2,3-trihydroxypropane (glycerol):

$$\begin{array}{r} CH_3(CH_2)_{16}COOCH_2 \\ | \\ CH_3(CH_2)_{16}COOCH \\ | \\ CH_3(CH_2)_{16}COOCH_2 \end{array} + 3NaOH \longrightarrow 3CH_3(CH_2)_{16}COONa + \begin{array}{l} CH_2OH \\ | \\ CHOH \\ | \\ CH_2OH \end{array}$$

Solvents

The smaller molecule esters are important solvents in, for example, the decaffeination of tea and coffee, and in paints and varnishes (such as nail varnish remover).

Polyesters

A **polyester** is a polymer or chain made up of repeating units joined by an ester linkage. Polymerisation was covered briefly in Chapter 14 where we covered polyalkenes. The polymerisation is different here because two different molecules are needed, rather than just one.

The most important commercial polyester is known as Terylene® and is produced by reacting a dicarboxylic acid with a dialcohol:

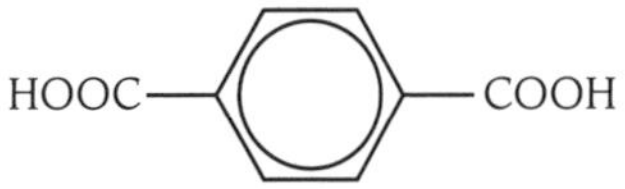

$$HOOC-C_6H_4-COOH \qquad HO-CH_2CH_2-OH$$

Benzene-1,4-dicarboxylic acid   Ethane-1,2-diol

When these molecules react together they form an ester linkage with the loss of a water molecule. The new molecule has a carboxylic acid group at one end and an alcohol group at the other. Therefore, it can continue to react, alternately adding a dialcohol and a dicarboxylic acid as shown below:

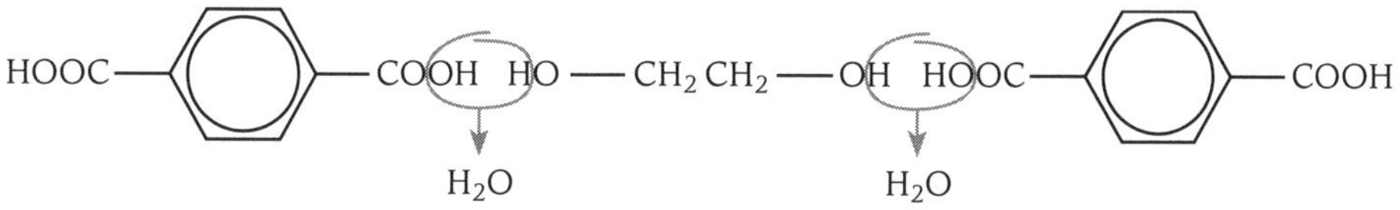

HOOC—C_6H_4—COOH HO—CH_2CH_2—OH HOOC—C_6H_4—COOH

H_2O H_2O

Polymers are covered in more detail in Chapter 19.

18 Nitrogen compounds

This chapter is needed only for the A2 examination.

Primary amines

Formation

Alkyl amines

The reaction to use for the formation of an alkyl amine is the reduction of a nitrile using lithium tetrahydridoaluminate(III) (lithium aluminium hydride). The reaction takes place in ethoxyethane solution and the amine is produced on adding a small amount of dilute acid. This does *not* work with $NaBH_4$:

$$CH_3CN + 4[H] \rightarrow CH_3CH_2NH_2$$

This reaction can also be achieved by heating the nitrile in hydrogen gas in the presence of a platinum, palladium or nickel catalyst:

$$CH_3CN + 2H_2 \rightarrow CH_3CH_2NH_2$$

Phenylamine

Phenylamine is prepared from nitrobenzene in a two-stage process. First, nitrobenzene is heated under reflux in a boiling water bath with a mixture of tin and concentrated hydrochloric acid. Because of the acidic conditions, rather than getting phenylamine directly, the nitrobenzene is reduced to phenylammonium ions. The lone pair on the nitrogen in the phenylamine picks up a hydrogen ion from the acid:

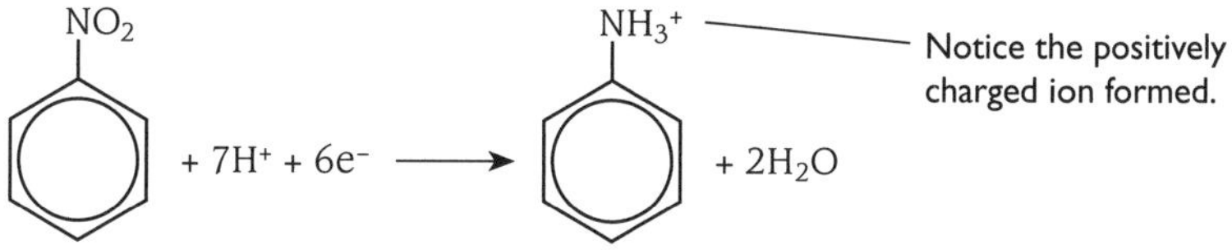

$$C_6H_5NO_2 + 7H^+ + 6e^- \longrightarrow C_6H_5NH_3^+ + 2H_2O$$

Notice the positively charged ion formed.

The second stage is to add sodium hydroxide solution to remove the hydrogen ion:

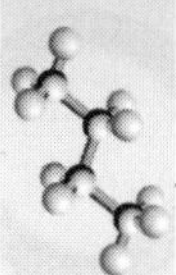

$$C_6H_5NH_3^+ + OH^- \longrightarrow C_6H_5NH_2 + H_2O$$

The phenylamine is then extracted by steam distillation.

Basicity

When considering the basic properties of alkyl amines, it is easiest to think of them as substituted ammonia molecules. If you remember that bases are proton acceptors it is not too difficult to make the comparison. Compare the two equations:

$NH_3 + H_3O^+ \rightarrow NH_4^+ + H_2O$

$CH_3CH_2NH_2 + H_3O^+ \rightarrow CH_3CH_2NH_3^+ + H_2O$

The same rule applies to secondary and tertiary amines, with alkyl groups substituting hydrogen atoms on the nitrogen atom:

$(CH_3CH_2)_2NH + H_3O^+ \rightarrow (CH_3CH_2)_2NH_2^+ + H_2O$

$(CH_3CH_2)_3N + H_3O^+ \rightarrow (CH_3CH_2)_3NH^+ + H_2O$

Amines are generally stronger bases than ammonia. This is because alkyl groups are electron donating, pushing negative charge onto the nitrogen atom and strengthening the attraction to the proton. The more alkyl groups there are attached to the nitrogen, the stronger the base formed.

By contrast, phenylamine is a much *weaker* base than ammonia. This is because the lone pair of electrons on the nitrogen atom is delocalised with the π-electrons on the benzene ring and is thus no longer as available to attract hydrogen ions.

Reactions of phenylamine

With aqueous bromine

With bromine water, phenylamine gives a fairly instant white precipitate of 2,4,6-tribromophenylamine at room temperature:

$$C_6H_5NH_2 + 3Br_2 \longrightarrow C_6H_2Br_3NH_2 + 3HBr$$

This is an example of electrophilic substitution of the benzene ring.

With nitrous acid

Phenylamine undergoes different reactions with nitrous acid depending on the temperature of the reaction mixture.

At low temperatures (<10 °C) phenylamine is dissolved in cold hydrochloric acid and then a solution of cold sodium nitrite is added slowly. The slow addition is so that the reaction mixture remains around 5 °C. A solution containing benzenediazonium ions is obtained:

$$C_6H_5NH_2 + HNO_2 + H^+ \longrightarrow C_6H_5\overset{+}{N}{\equiv}N + 2H_2O$$

Note that when drawing the structure of a benzenediazonium salt, the positive charge is shown on the nitrogen atom closest to the benzene ring.

The benzenediazonium salt is not isolated, but is often reacted by adding a cold solution of phenol in sodium hydroxide to the mixture:

$$C_6H_5\overset{+}{N}{\equiv}N + C_6H_5O^- \longrightarrow C_6H_5{-}N{=}N{-}C_6H_4{-}OH$$

The result is a yellow, orange or red dye, the colour of which depends on the nature of the phenol used. With naphthalen-2-ol (2-naphthol), the following compound is formed:

$$C_6H_5\overset{+}{N}{\equiv}N + C_{10}H_7O^- \longrightarrow C_6H_5{-}N{=}N{-}C_{10}H_6{-}OH$$

At temperatures above 10 °C, a reaction takes place that produces a complex organic mixture containing mainly phenol. Nitrogen gas is evolved:

$$C_6H_5NH_2 + HNO_2 \rightarrow C_6H_5OH + N_2 + H_2O$$

Try this yourself

(38) Explain why phenylamine is a much weaker base than ethylamine.

Amides

Formation from acyl halides

We saw in Chapter 17 how acyl halides react with primary amines. They also react with ammonia to form amides. This is normally carried out by adding the acyl halide to a concentrated solution of ammonia. The reaction is violent, producing clouds of ammonium chloride:

$$CH_3COCl + 2NH_3 \rightarrow CH_3CONH_2 + NH_4Cl$$

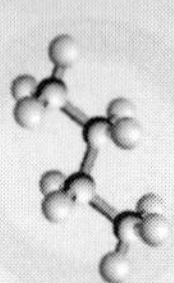

Hydrolysis

Amides are hydrolysed by heating with aqueous acids or aqueous alkalis:

$$CH_3CONH_2 + HCl + H_2O \rightarrow CH_3CO_2H + NH_4^+Cl^-$$

$$CH_3CONH_2 + NaOH \rightarrow CH_3CO_2^-Na^+ + NH_3$$

The second reaction is sometimes used as a test for amides, because the ammonia produced is easily detected.

Formation of polyamides

Polyamides are a very important group of polymers that include nylon and Kevlar®. The polymerisation reaction involves two monomers that react by condensation to form an amide bond between them:

$$-\overset{\overset{\displaystyle O}{\|}}{C}-\underset{\underset{\displaystyle H}{|}}{N}-$$

This means that the monomers have to be diamines and dicarboxylic acids.

Nylon-6,6

Nylon is an important polymer used as both a fibre and a bulk solid. It was used originally as a replacement for silk, but was soon being used in strings for musical instruments and in rope. In its solid form it has become a replacement for metal in low-medium stress applications, such as gears. For nylon, the monomers are hexanedioic acid and 1,6-diaminohexane:

$HO_2C(CH_2)_4CO_2H$	$H_2N(CH_2)_6NH_2$
Hexanedioic acid	1,6-diaminohexane

Each of these molecules has six carbon atoms, so this form of nylon is called nylon-6,6. Water is lost in the formation of each amide bond:

$$HO-\overset{\overset{\displaystyle O}{\|}}{C}-(CH_2)_4-\overset{\overset{\displaystyle O}{\|}}{C}-OH \quad H-\underset{\underset{\displaystyle H}{|}}{N}-(CH_2)_6-\underset{\underset{\displaystyle H}{|}}{N}-H \quad \rightarrow H_2O \quad H_2O \quad H_2O$$

$$-\overset{\overset{\displaystyle O}{\|}}{C}-(CH_2)_4-\overset{\overset{\displaystyle O}{\|}}{C}-\underset{\underset{\displaystyle H}{|}}{N}-(CH_2)_6-\underset{\underset{\displaystyle H}{|}}{N}-\overset{\overset{\displaystyle O}{\|}}{C}-(CH_2)_4-\overset{\overset{\displaystyle O}{\|}}{C}-\underset{\underset{\displaystyle H}{|}}{N}-(CH_2)_6-\underset{\underset{\displaystyle H}{|}}{N}-$$

Nylon-6

Another form of nylon can be produced from a single monomer called caprolactam:

Caprolactam

At first sight, caprolactam does not seem to be a very promising monomer because it exists as a ring. However, when caprolactam is heated in the presence of about 10% water at 250 °C, the ring is broken between the carbon and the nitrogen and an amino acid is formed. The amino acid is then able to polymerise by a condensation reaction with the elimination of water to form nylon-6.

$$-NH-(CH_2)_5-CO-NH-(CH_2)_5-CO-NH-(CH_2)_5-CO-$$

Kevlar®

Kevlar® is a relatively new polymer with exceptional strength-to-weight properties. It is used in body armour such as bulletproof vests. It is also used in sporting applications such as bicycle tyres, sails and racing cars.

The monomers for Kevlar® are the aryl equivalents of those used for nylon:

$HOOC-C_6H_4-COOH$

Benzene-1,4-dicarboxylic acid

$H_2N-C_6H_4-NH_2$

1,4-diaminobenzene

A similar condensation reaction takes place:

$$-CO-C_6H_4-CO-NH-C_6H_4-NH-CO-C_6H_4-CO-NH-C_6H_4-NH-$$

One disadvantage of polyamides is that they are hydrolysed by strong acids (although Kevlar® is more resistant).

Try this yourself

(39) Describe two main differences between *condensation* polymerisation as described here and *addition* polymerisation as described in Chapter 19.

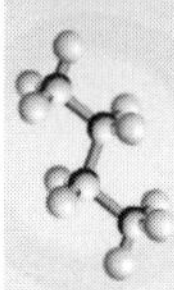

Amino acids and proteins

Acid-base properties of amino acids

As the name suggests, amino acids are organic compounds that contain both a carboxylic acid group and an amine group. The amino acids that are important biologically have the amine group attached to the *same* carbon atom as the carboxylic acid group. They are known as 2-aminoacids (or alpha-amino acids). Glycine (2-aminoethanoic acid) is the simplest:

$H_2N—CH_2—CO_2H$
Glycine (2-aminoethanoic acid)

Amino acids are crystalline, high melting point (>200 °C) solids. Such high melting points are unusual for a substance with molecules of this size; they are a result of internal ionisation. Even in the solid state, amino acids exist as **zwitterions** in which a proton has been lost from the carboxyl group and accepted by the nitrogen of the amine group:

$H_3N^+—CH_2—CO_2^-$
Zwitterion of glycine

Hence, instead of hydrogen bonds between the amino acid molecules there are stronger ionic (electrovalent) bonds. This is reflected in the relative lack of solubility of amino acids in non-aqueous solvents compared with their solubility in water.

The zwitterions exhibit acid–base behaviour since they can accept and donate protons.

In acid, a proton is accepted onto the carboxylic acid anion forming a unit with an overall positive charge:

$$H_3N^+—CH_2—CO_2^- + H_3O^+ \rightarrow H_3N^+—CH_2—CO_2H + H_2O$$

In alkali, the reverse occurs with the loss of a proton from the nitrogen atom:

$$H_3N^+—CH_2—CO_2^- + OH^- \rightarrow H_2N—CH_2—CO_2^- + H_2O$$

The species present in a given solution depends on the pH of the solution.

Formation of peptide bonds

If we consider amino acids in their non-ionic form, it is easy to see that they have the potential to react together to form a polymer (in a similar way to nylon). The carboxylic acid and amine groups will react to form an amide linkage (this is called a **peptide bond** in biological systems) with the elimination of a water molecule. This reaction is of immense biological significance because the polymers formed are polypeptides or proteins, and form part of the chemistry of all living organisms.

There are 20 amino acids of significance in biological systems and since each possesses a carboxylic acid and an amine group, the possibilities for constructing polypeptides and proteins are enormous.

Consider two simple amino acids, glycine (2-aminoethanoic acid) and alanine (2-aminopropanoic acid). These can be joined in two ways:

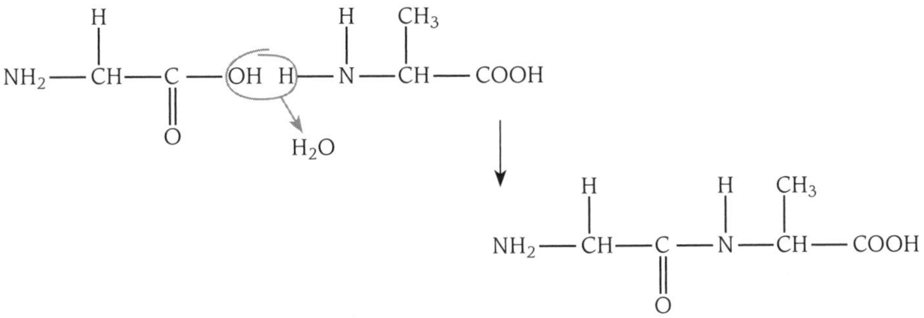

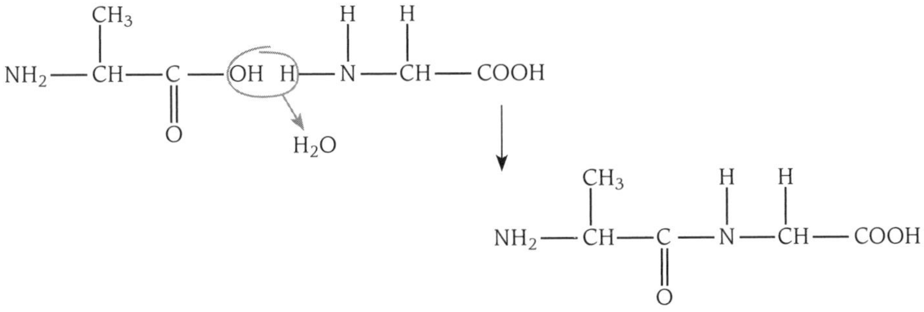

$$NH_2-CH(H)-C(=O)-OH \quad H-N(H)-CH(CH_3)-COOH \rightarrow H_2O$$

$$\downarrow$$

$$NH_2-CH(H)-C(=O)-N(H)-CH(CH_3)-COOH$$

$$NH_2-CH(CH_3)-C(=O)-OH \quad H-N(H)-CH(H)-COOH \rightarrow H_2O$$

$$\downarrow$$

$$NH_2-CH(CH_3)-C(=O)-N(H)-CH(H)-COOH$$

The two dipeptides formed have different structures. A typical protein is formed from between 50 and 200 amino acids joined in a variety of sequences, so you can see how complex protein chemistry can be!

Hydrolysis of proteins

Because proteins contain amide (or peptide) bonds, they can be hydrolysed. This can help us to analyse the amino acids that make up a given protein (or polypeptide).

The traditional way of doing this was to heat the protein in 6 mol dm^{-3} hydrochloric acid at around 100 °C for 24 hours. It has recently been discovered that this process can be speeded up considerably. The protein is now placed in 6 mol dm^{-3} hydrochloric acid under an atmosphere of nitrogen and microwaved for up to 30 minutes.

The amide (peptide) bonds are broken leaving the amino acids in their protonated form, for example:

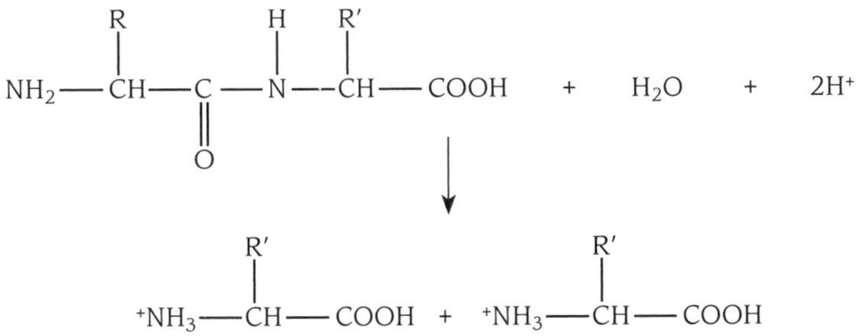

$$NH_2-CH(R)-C(=O)-N(H)-CH(R')-COOH \quad + \quad H_2O \quad + \quad 2H^+$$

$$\downarrow$$

$$^+NH_3-CH(R')-COOH \;+\; ^+NH_3-CH(R')-COOH$$

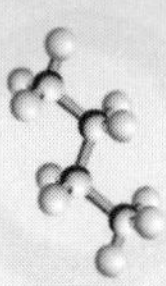

19 Polymerisation

This chapter is needed for both the AS and A2 examinations. Addition polymerisation may be tested at AS; condensation polymerisation is needed for A2.

Addition polymerisation

Poly(ethene)

Addition polymerisation takes place when molecules containing a C=C double bond are joined together to form a long chain. The simplest of these reactions is the polymerisation of ethene:

$$n\mathrm{CH_2{=}CH_2} \rightarrow (\mathrm{{-}CH_2{-}CH_2{-}})_n$$

In this reaction, the conditions needed are a temperature of around 200 °C, a pressure of around 2000 atmospheres and a small amount of oxygen to initiate the reaction. Under these conditions the chain could be made up of anything between 2000 and 20 000 molecules. This reaction produces a form of poly(ethene) called **low-density poly(ethene)** due to the fact that the chains are branching and do not allow close packing. This polymer is used to make plastic bags and for other low-strength sheet applications.

Another form of poly(ethene), **high-density poly(ethene)**, can be produced but the conditions used are very different. The temperature is much lower at around 60 °C, the pressure is only a few atmospheres and a catalyst is required. These conditions cause the chains to grow in a much more ordered way enabling them to pack together much more closely, hence increasing the density of the bulk polymer. This polymer is used to make plastic containers, washing-up bowls and some plastic pipes.

Poly(chloroethene)

Chloroethene is similar to ethene, but has one hydrogen atom replaced by a chlorine atom. The double bond is still present and so the molecule can be polymerised:

$$n\mathrm{CH_2{=}CH}Cl \rightarrow (\mathrm{{-}CH_2{-}CH}Cl{-})_n$$

This polymer is still called after the old name for chloroethene — vinyl chloride — so you will see it referred to as polyvinylchloride or PVC. Although the reaction is the same, it is usual to draw the molecules showing the chlorines to one side of the chain as below:

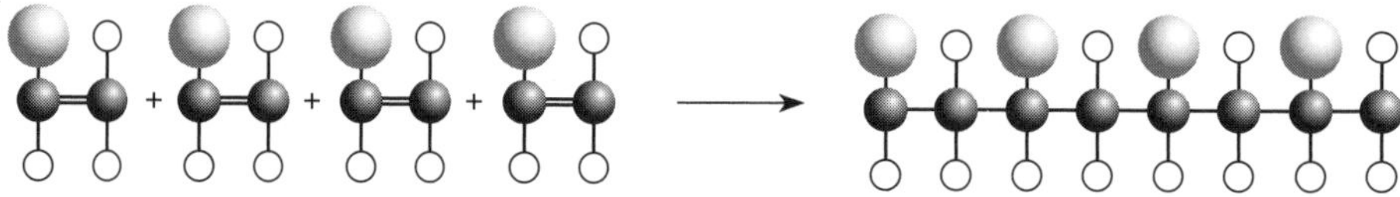

The equation to show this is as follows:

$$n\,\mathrm{C(Cl)H{=}CH_2} \longrightarrow \mathrm{-\!\left[C(Cl)H-CH_2\right]_{\mathit{n}}\!-}$$

Poly(chloroethene) is used to make a wide range of products including guttering and plastic window frames. It can be rather hard and rigid. Chemicals called plasticisers can be added to increase the flexibility. This increases the range of uses — for example, electrical cable insulation, sheet materials for flooring, footwear and clothing.

Disposal of polyalkenes

The use of polyalkenes has become so widespread that we are now faced with the disposal of thousands of tonnes of waste polyalkenes each year. This causes a number of problems:

- In general, polyalkenes are not biodegradable, so burying them in landfill sites is useless. Plastic bags and bottles can prove harmful to wildlife. Even fish have been found to have tiny particles of plastic in their stomachs. New biodegradable polymers have been developed for use in some types of packaging, such as plastic bags.
- There is a large amount of hydrocarbons locked in the polyalkenes that are thrown away. This is wasting a precious resource — crude oil. In some parts of the world, some plastics are sorted and some used to fuel power generation plants. It is important that this use is seen as a replacement for oil because simply incinerating the polyalkenes to dispose of them would add to the carbon dioxide put into the atmosphere.
- There are so many different types of polymer used that it is not easy to sort them quickly for future use, either for recycling or as a fuel. Indeed some polyalkenes, such as PVC, produce toxic gases such as hydrogen chloride and dioxins when they are burned.

Condensation polymerisation

This section is only tested in the A2 examination.

Characteristics of condensation polymerisation

The points to remember about condensation polymerisation are as follows:

- A small molecule, such as water, is eliminated as each monomer is added.
- You need two different monomers (usually but *not* always, e.g. nylon-6)
- The properties of the polymer depend on the monomer molecules used.

Polyamides

We saw in Chapter 18 how the polyamides nylon-6, nylon-6,6 and polypeptides and proteins are formed by condensation polymerisation. There is nothing new to learn

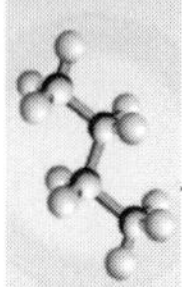

here — simply remember that in condensation polymerisation small molecules (often, but not necessarily, water) are eliminated as the polymer grows.

Polyesters

We saw in Chapter 17 how another group of condensation polymers — polyesters in the form of Terylene® — could be formed by condensation polymerisation. There is nothing new to learn here.

Identifying polymers

The last part of this section of the syllabus outlines what you might be asked to deduce about a given polymer. You may be asked to:

- predict the type of polymerisation reaction for a given monomer or pair of monomers
- deduce the repeat unit of a polymer obtained from a given monomer or pair of monomers
- deduce the type of polymerisation reaction that produces a given section of a polymer molecule
- identify the monomer(s) present in a given section of a polymer molecule

It is perhaps easiest to see what you need to be able to do by looking at some examples. Let's look at each of the four points in turn.

Example 1

Predict what type of polymerisation would take place with each of the following monomers, explaining your answers:

(a) ethene and propene

(b) 1,2-dihydroxyethane and 1,4-benzenedicarboxylic acid

(c) aminoethanoic acid (glycine)

Answer

(a) Both ethene and propene contain double bonds, so they react by addition polymerisation to produce a mixed polyalkene.

(b) The key here is recognising that there is a hydroxy compound and a carboxylic acid and remembering that they react to form an ester. Each monomer has two identical functional groups, so together they can form a polymer. When a carboxylic acid reacts with a hydroxyl compound to form an ester, water is eliminated. Therefore, this is condensation polymerisation.

(c) Here there is only one monomer. It has an amine group at one end and a carboxylic acid group at the other. These two groups (on different molecules) react together to form an amide, and since the monomer has a functional group at each end, polymerisation can take place. When an amide is formed a condensation reaction takes place and water is eliminated, so this is condensation polymerisation.

Example 2

Draw the repeat unit for the polymer formed when the following monomers react together:

(a) 1,6-diaminohexane, $H_2N(CH_2)_6NH_2$, and hexanedioic acid, $HO_2C(CH_2)_4CO_2H$

(b) caprolactam

O, H, C—N, CH_2, CH_2, H_2C, CH_2, CH_2

Answer

(a) In order to tackle these questions you must think about the way the monomers combine. In this case they react with the elimination of water:

HO—C(=O)—$(CH_2)_4$—C(=O)—OH H—N(H)—$(CH_2)_6$—N(H)—H

H_2O H_2O H_2O

—$(CH_2)_4$—C(=O)—N(H)—$(CH_2)_6$—N(H)—C(=O)—$(CH_2)_4$—C(=O)—N(H)—$(CH_2)_6$—

Once you have drawn the polymer, the repeat unit is formed from each *pair* of monomers. In this case, it is the part indicated between the lines:

| C(=O)—$(CH_2)_4$—C(=O)—N(H)—$(CH_2)_6$—N(H) | C(=O)—$(CH_2)_4$—C(=O)—N(H)—$(CH_2)_6$—N(H)—

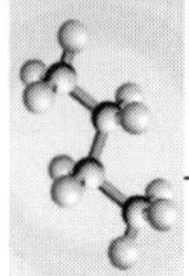

(b) The first thing to notice here is that, as it stands, caprolactam is not really a monomer. To become a monomer, the ring has to break between the carbon and nitrogen. The molecule formed is:

These molecules then polymerise. Starting from only one monomer, the repeat unit has to include most of it! The repeat unit is indicated between the lines:

Example 3

The diagrams below show parts of two polymers:

(a)

(b)

Deduce what type of polymerisation produced each polymer, explaining your answers.

Answer

(a) In this polymer the bonds within the chain are all carbon–carbon single bonds. This can only have resulted from addition polymerisation, because in condensation polymerisation there are *always* other atoms in the polymer chain.

(b) This polymer contains an ester linkage denoted by the —O—C═O group. This can only have been formed from two monomers by condensation polymerisation with the loss of water.

Example 4

The diagram shows a length of polymer containing the residues of a monomer(s). Draw the structure of the original monomer(s).

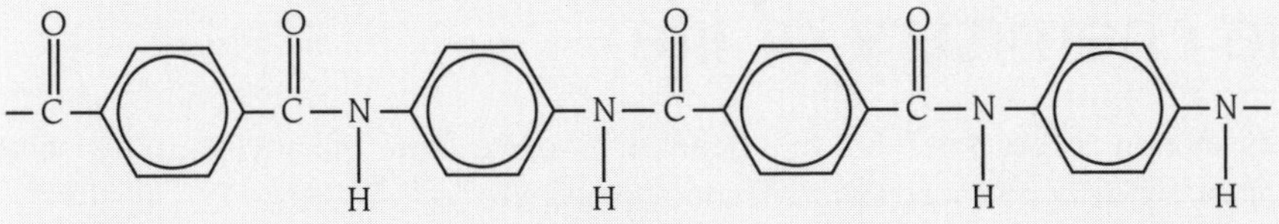

Answer

When tackling this sort of question you need to work out where the monomer residue starts and ends, and to decide whether the polymer was made by addition or condensation polymerisation.

If you think it was made by addition, then the monomer will contain a C═C double bond. If you think it was made by condensation polymerisation you need to think about the functional groups that were at either end of the monomer.

The structure of the polymer shows two types of monomer, as shown by the lines in the diagram below:

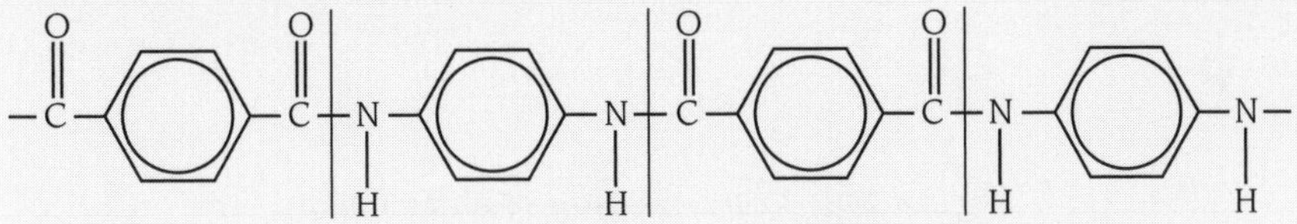

You can also identify an amide linkage, which should tell you that the monomers were probably a dicarboxylic acid and a diamine:

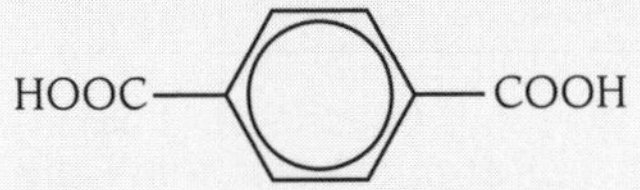

HOOC—C_6H_4—COOH

Benzene-1,4-dicarboxylic acid

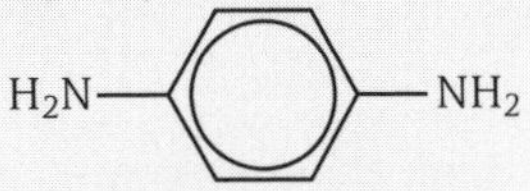

H_2N—C_6H_4—NH_2

1,4-diaminobenzene

20 Applications of chemistry

This chapter is only needed for the A2 examination. It is important to realise that although the three sections of this part of the syllabus have both content and learning outcomes, the purpose of this chapter is for students to apply their knowledge of chemistry in new contexts. Here the contexts are spelled out in terms of areas of chemistry in which modern developments are important.

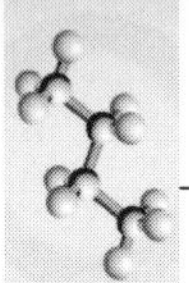

There may be questions in the exam paper where the specific application has not been covered in the syllabus, but the chemistry behind it has. You need to be able to recognise and apply this chemistry in the context of the question.

The chemistry of life

This section covers a number of biochemical topics. Note the advice in the syllabus that you will *not* be asked to recall specific proteins.

Protein chemistry

The first part of this section is based on work covered in Chapter 18. You should know that proteins are polymers formed by the condensation of amino acids (Figure 20.1). You also need to remember that there are 20 different biologically active amino acids.

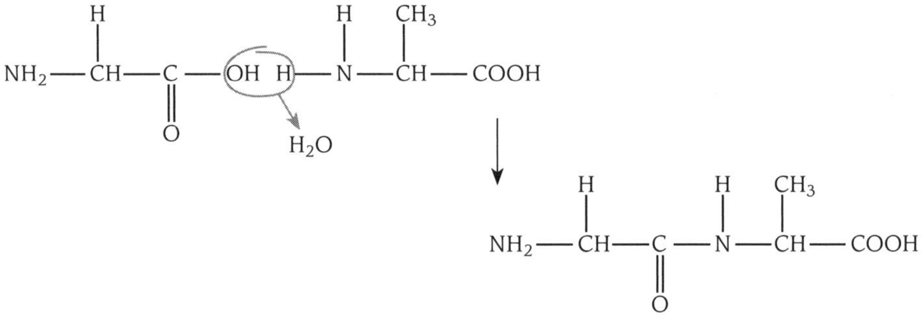

$$NH_2{-}CH(H){-}C({=}O){-}OH \;\; H{-}N(H){-}CH(CH_3){-}COOH \xrightarrow{-H_2O} NH_2{-}CH(H){-}C({=}O){-}N(H){-}CH(CH_3){-}COOH$$

Figure 20.1 Condensation of amino acids

Figure 20.1 is a reminder of how amino acids react together. You need to remember that there may be up to 200 amino acids in a protein chain, and that they can be arranged in any sequence using any of the 20 different 2-amino acids (alpha-amino acids).

If you think about a polypeptide containing just four amino acids, there are lots of ways of putting them in a sequence. Even if the possibility of amino acid repeats — for example AABC — is ignored, there are still a lot of combinations. How many do you think there are?

ABCD, ABDC, ADBC, ACBD... etc.

The sequence of amino acids is called the **primary structure** of a protein. Polypeptides have a 'backbone' to their structure that runs the length of the chain. If you look at Figure 20.1 you can see that the only difference from one amino acid to the next is in the side chains, the backbone is the same and has the pattern —C—C—N—C—C—N— etc. The chain can be rotated about carbon-to-carbon bonds to make it twist or fold, and this gives the protein its **secondary structure.** Hydrogen bonds between peptide bonds in different sections of the chain lead to two main types of structure — the α-helix (Figure 20.2(a)) and the β-pleated sheet (Figure 20.2(b)).

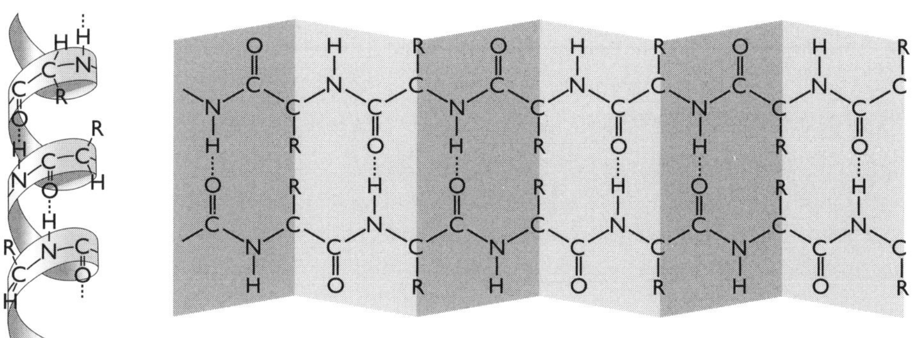

Figure 20.2 Secondary structure of a protein (a) α-helix (b) β-pleated sheet

The sequence of amino acids is important because the groups that form side chains in a polypeptide or protein can interact with other side chains further along the chain as it bends. This is known as the **tertiary structure** of a protein. Some examples of side chains of the different amino acids and the bonds that can form between them are shown in Table 20.1.

Table 20.1

Amino acid	Structure	Type of bond formed
Alanine	$H_2NC(CH_3)HCO_2H$	van der Waals
Serine	$H_2NC(CH_2OH)HCO_2H$	Hydrogen bonds
Aspartic acid	$H_2NC(CH_2CO_2H)HCO_2H$	Ionic (electrovalent)
Lysine	$H_2NC((CH_2)_4NH_2)HCO_2H$	Ionic (electrovalent)
Cysteine	$H_2NC(CH_2SH)HCO_2H$	Disulfide bridges

Things to remember

- Proteins are polymers formed from 2-amino acids (alpha-amino acids) by condensation polymerisation.
- 2-amino acids have different side-chains, which gives them their polar or non-polar nature; 20 different 2-amino acids are used in the body.
- 2-amino acids exist as **zwitterions** in both the solid state and in aqueous solution.
- The **primary structure** of a protein is the sequence of 2-amino acids in the polypeptide chain.
- The polypeptide chain has direction, with one end (N-terminal) an amine group and the other end (C-terminal) a carboxylic acid.
- The **secondary structure** of a protein results from hydrogen bonds formed between different parts of the folded chain. The most stable of these structures are in the form of α-helices and β-pleated sheets.
- The **tertiary structure** of a protein arises from the folding of the polypeptide chain due to interactions between the side chains of the amino acids. These interactions include van der Waals forces, hydrogen bonds and ionic (electrovalent) bonds, together with covalent disulfide bridges.

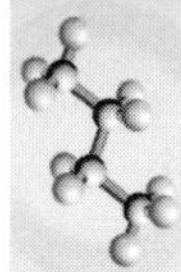

- The functioning of a protein is linked directly to its three-dimensional structure.
- Proteins can be hydrolysed back to amino acids (see Chapter 18).

Proteins as enzymes

Enzymes are particular types of protein that behave as catalysts enabling chemical reactions to take place efficiently at body temperature. Unlike inorganic catalysts that are often able to catalyse a range of reactions with similar substances, enzymes are specific in their behaviour, generally catalysing one particular reaction.

Most enzymes are water-soluble (the body is an aqueous environment), globular (ball-shaped) proteins. The complex folding that gives the tertiary structure to these proteins creates channels and grooves in the surface of the enzyme. It is the precise shape of these features that fit a given substrate molecule when it reacts. This is known as the **active site** of the enzyme.

You may be asked to draw a diagram to help explain this, so think about the important parts of the process (Figure 20.3).

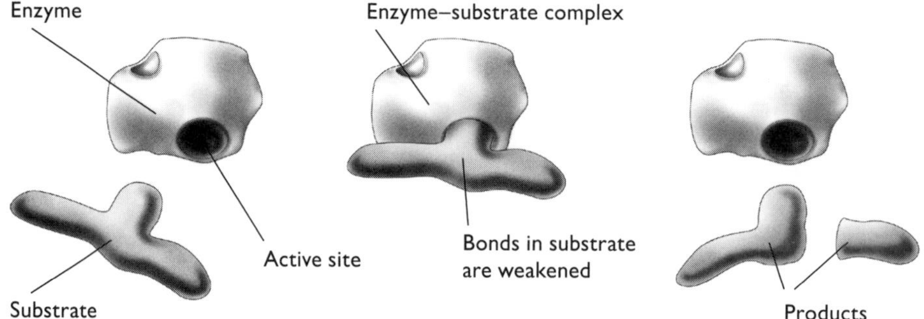

Figure 20.3 Reaction between an enzyme molecule and a substrate molecule

Enzymes, like other catalysts, work by providing an alternative reaction pathway of lower activation energy, E_{cat}, as shown in Figure 20.4.

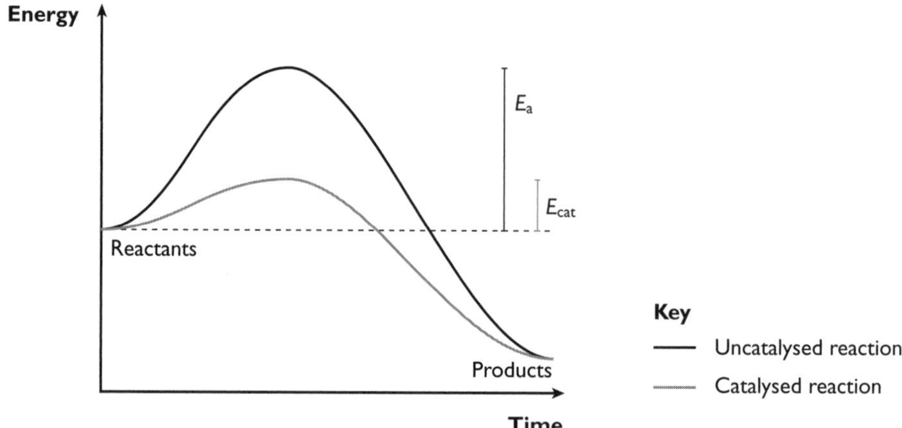

Figure 20.4 A reaction with and without an enzyme catalyst

The catalysed reaction requires less energy than the uncatalysed reaction. It is usual to describe the reaction as follows:

enzyme + substrate ⇌ enzyme–substrate complex → enzyme + products

The first step is reversible because unless the available energy is greater than E_{cat}, the complex may break down with no product being formed. Once the products have been formed they leave the active site of the enzyme, making way for a new substrate molecule to bind.

Factors affecting enzyme activity

The activity of an enzyme in catalysing a given reaction is dependent on the weak interactions that maintain the tertiary structure of the protein. As a result, even small changes in conditions can disrupt enzyme activity. The two most obvious changes are the temperature and pH of the solution in which the reaction takes place.

The effect of temperature

Temperature is the more complex of the two factors because it influences:

- the speed of the molecules
- the thermal stability of the enzyme and of the substrate
- the activation energy, E_{cat}, of the catalysed reaction

Given that most enzyme-catalysed reactions occur at temperatures close to Earth's ambient temperature, we do not expect a significant reaction rate at around 0 °C (in aqueous solutions this would be approaching freezing point). High temperatures disrupt the relatively weak bonds that maintain the tertiary structure of the protein. This leaves a range of approximately 0 °C to 60 °C for enzyme-catalysed reactions with a peak of activity around 40 °C (Figure 20.5).

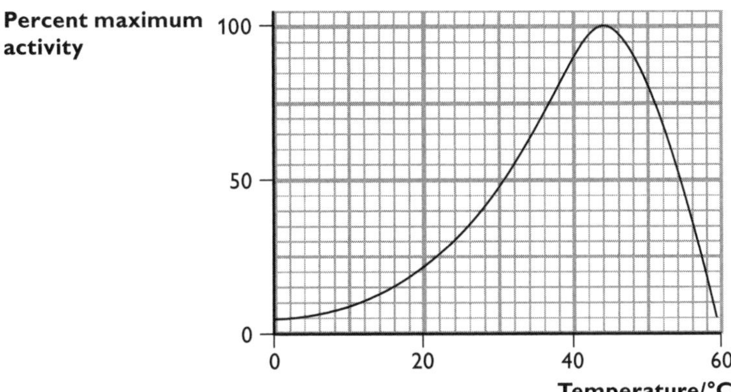

Figure 20.5 The effect of temperature on enzyme activity

One of the reasons why a high fever can be so dangerous is that at temperatures much above 40 °C some enzyme-catalysed processes no longer take place.

The effect of changing pH

Enzyme activity is dependent on the pH of the environment for two main reasons:

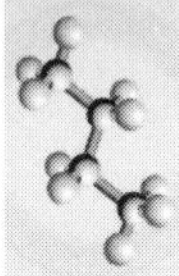

- Small changes in pH can affect the ionisation of amino acid side-chains, which can alter the ability of a substrate to bind to the active site.
- Extremes of pH can cause proteins to denature (lose the three-dimensional arrangement of the protein chains), which can also affect their solubility.

Most enzymes operate over a fairly narrow pH range around an optimum pH. These can be quite different, depending on the enzyme (Figure 20.6).

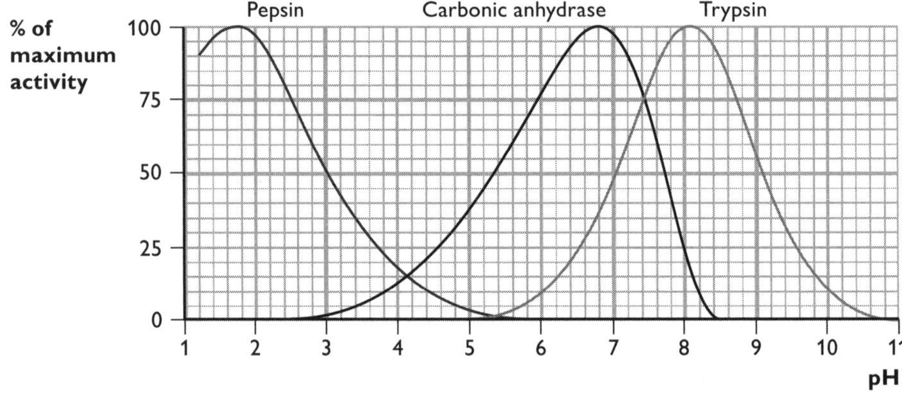

***Figure 20.6** The effect of pH on the activity of three different enzymes*

As you can see in Figure 20.6, pepsin, which acts on proteins in the acid conditions of the stomach, has an optimum pH of around 2; carbonic anhydrase, which operates in red blood cells, has an optimum pH of around 7; trypsin, which acts on peptides in the slightly alkaline conditions of the small intestine, has an optimum pH of around 8.

Inhibition of enzyme activity

As well as changes in the structure of the enzyme brought about by changes in temperature or pH, enzyme activity can be affected by the presence of other molecules. These molecules are known as **inhibitors** and can affect the enzyme in one of two ways.

Competitive inhibition

Enzyme activity may be reduced in the presence of a molecule that has a similar structure to the substrate. This enables it to bind to the active site of the enzyme, competing with the substrate molecule. The result of such competition on the effectiveness of the enzyme depends on the relative concentrations of the substrate and inhibitor molecules. Unlike the changes in temperature or pH, the ability of the enzyme to function is not changed, it's just that the active site can be blocked by the inhibitor (Figure 20.7).

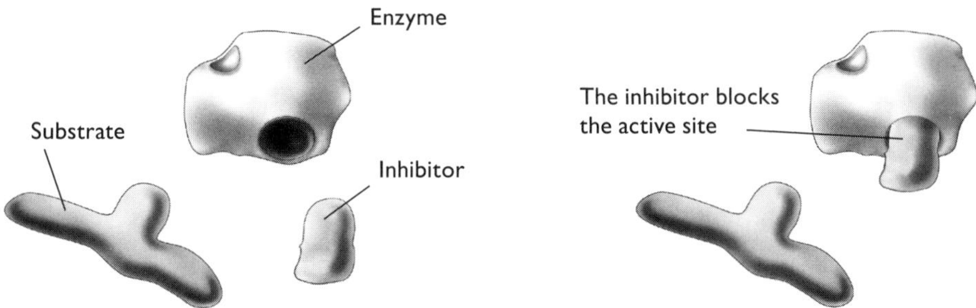

Figure 20.7 Competitive inhibition

The extent to which competitive inhibition affects the rate of the enzyme-catalysed reaction depends on:

- the concentration of substrate
- the concentration of inhibitor
- the relative strengths of bonds formed between the active site and the substrate, and active site and the inhibitor

The effects of this are shown in Figure 20.8.

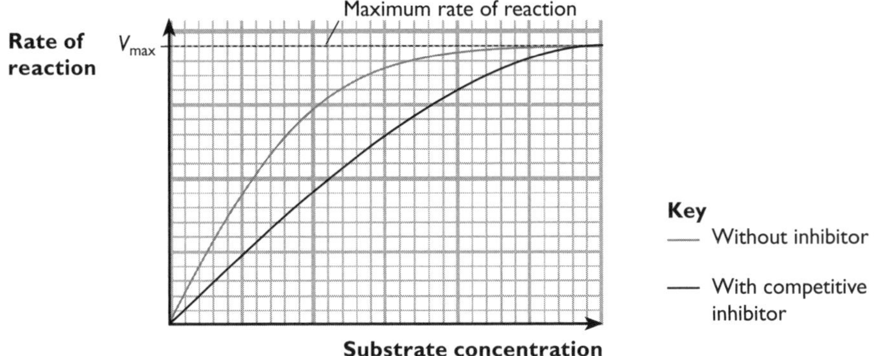

Figure 20.8 The effect of a competitive inhibitor on the rate of an enzyme-catalysed reaction

If there is a high enough substrate concentration it is possible to reach the maximum rate, V_{max}.

Non-competitive inhibition of enzyme activity

The activity of an enzyme can also be reduced, or even stopped, by a molecule binding to an area other than the active site. This can change the shape of the active site making it impossible for the substrate to bind to it (Figure 20.9). This type of inhibitor molecule does not need to resemble the substrate and can be quite small.

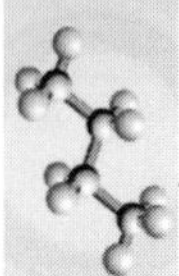

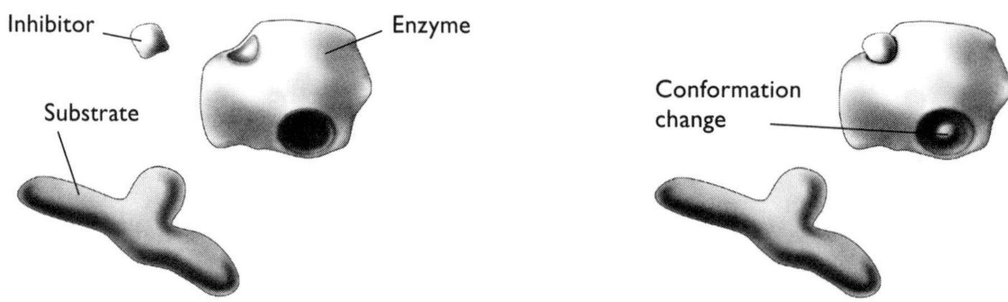

Figure 20.9 Non-competitive inhibition

The extent to which non-competitive inhibition affects the rate of the enzyme-catalysed reaction depends on:

- the concentration of the inhibitor
- the affinity of the enzyme for the inhibitor

The effects are shown in Figure 20.10, which is a graph of rate of reaction against substrate concentration. Competitive inhibition is included in Figure 20.10 for comparative purposes. K_m is the concentration of substrate at which the rate has reached half the maximum rate, V_{max}.

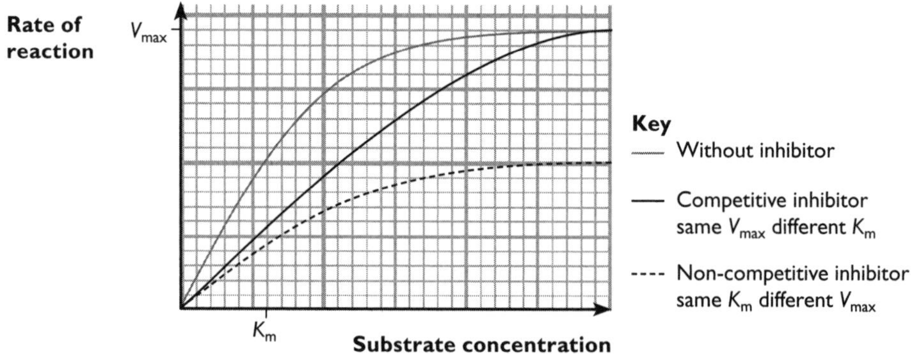

Figure 20.10 Comparison of the effects of non-competitive and competitive inhibitors on the rate of an enzyme-catalysed reaction

Cofactors

Many enzymes need an ion or a group to be present in order for them to work effectively. Some of these, such as metal ions — for example Zn^{2+} in carbonic anhydrase — are built into the structure of the enzyme itself. Others might be large organic molecules, often provided by a vitamin. These bind temporarily to the enzyme and assist with the transfer of groups or electrons not available within the active site itself (see p. 191).

Things to remember

- Enzymes act as biological catalysts by providing an alternative pathway with lower activation energy, E_{cat}, than the uncatalysed reaction.
- Enzymes are proteins that catalyse reactions in aqueous solutions at low temperatures and a specific pH.

- As a catalyst, an enzyme works with a specific substrate (unlike inorganic catalysts).
- The ability of an enzyme to function is dependent on its three-dimensional shape, specifically the precise shape of the active site.
- The substrate not only fits the active site, it binds to it to enable the reaction to take place.
- Enzyme-catalysed reactions may be reduced in their effectiveness by two types of inhibition:
  - in competitive inhibition, the inhibitor has a similar shape to the substrate and competes with it to occupy the active site
  - in non-competitive inhibition, the inhibitor does not bind to the active site, but binds elsewhere on the enzyme. This changes the shape of the enzyme and prevents the normal catalysed reaction from taking place.

Genetic information

You need to know about the role that DNA plays in cells, particularly with respect to protein synthesis. Some detailed chemistry of the processes within the cell is needed.

The structure of DNA

DNA (deoxyribonucleic acid) controls the passing of genetic information from one generation to the next. It also plays an important role in the synthesis of proteins.

The structure of DNA was discovered by Watson and Crick in 1953. It has enabled our understanding of heredity, plant and animal breeding, genetic diseases and the identification of individuals by their DNA 'fingerprint'.

DNA consists of a double strand of a macromolecule formed by condensation polymerisation. The monomers in the chain are called **nucleotides.** They are assembled into the chain by linking sugar (deoxyribose) and phosphate groups to form the backbone with bases attached as side chains. You can see this in block diagram form (the only form you will be asked to deal with in the exam) in Figure 20.11.

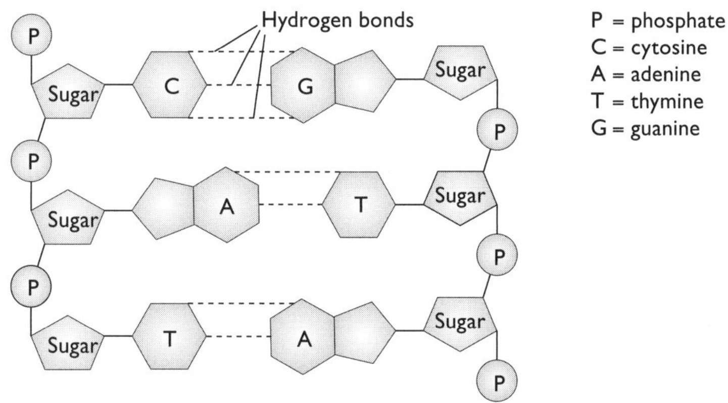

Figure 20.11 Block diagram to illustrate DNA structure

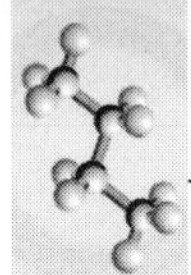

The sugar molecule is in the form of a five-membered ring and is joined to the phosphate group via an ester link. It is also joined to one of four bases:

- adenine (A)
- cytosine (C)
- guanine (G)
- thymine (T)

Within the double strand of DNA the bases are always paired in the same way — adenine with thymine and cytosine with guanine.

Tip A way to remember the base pairs is: All Things Come Good.

You do not need to learn the structures of the bases, which are complex. If you need more detail it will be given in the question.

DNA consists of two strands each made up as described, but running in opposite directions (anti-parallel). The strands coil around one another to form a double helix which is held together by hydrogen bonds between the pairs of bases. Two hydrogen bonds form between adenine and thymine, and three hydrogen bonds form between cytosine and guanine. The sugar–phosphate backbone is on the outside of the double helix, with the bases in the middle, as shown in Figure 20.12.

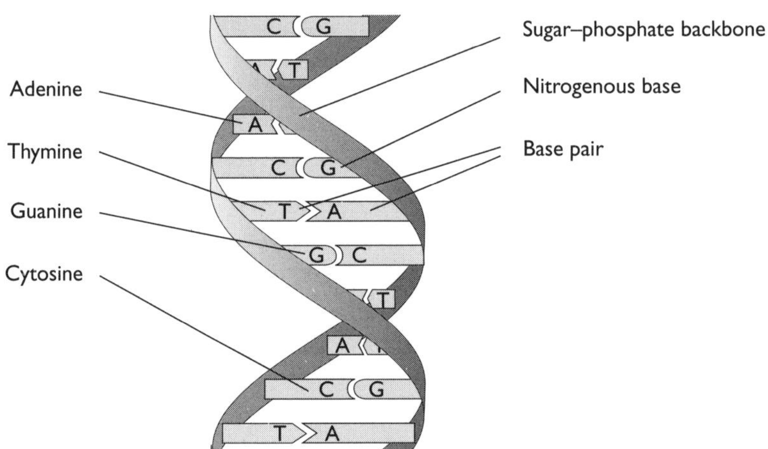

Figure 20.12 The structure of DNA

It is important to remember that the DNA in a cell codes for lots of different proteins. The section of DNA that codes for a particular protein is called a **gene.**

The structure of RNA

There is another nucleic acid, RNA, which is found in cells. The structure of RNA is similar to that of DNA, with a number of significant differences (Table 20.2).

Table 20.2

Factor	Deoxyribonucleic acid, DNA	Ribonucleic acid, RNA
Sugar	Pentose sugar is deoxyribose	Pentose sugar is ribose
Bases	Adenine Cytosine Guanine Thymine	Adenine Cytosine Guanine Uracil
Structure	Double helix with two anti-parallel strands	Single strand that can be folded to produce helical sections

There are three different types of RNA, each with a role to play in protein synthesis.

Messenger RNA (mRNA)

mRNA is complementary to a portion of one of the DNA strands and is copied from the DNA gene sequence for a particular polypeptide chain. The 'message' encoded in the mRNA is translated into the sequence of amino acids in the polypeptide. The mRNA travels out of the cell nucleus into the cytoplasm of the cell where it bonds to a ribosome.

Transfer RNA (tRNA)

Transfer RNA is a much smaller molecule than mRNA. Each tRNA molecule acts as a 'carrier' for an amino acid to a ribosome during protein synthesis. Each tRNA recognises the coding sequence for a particular amino acid in the mRNA.

Ribosomal RNA (rRNA)

There are a number of different rRNAs that form the major structural components of ribosomes. Ribosomes are found in the cell cytoplasm and are where protein synthesis occurs. It is the largest of the three forms of RNA with an M_r of up to 1 000 000.

The relationships between the different types of RNA can be seen in Figure 20.13.

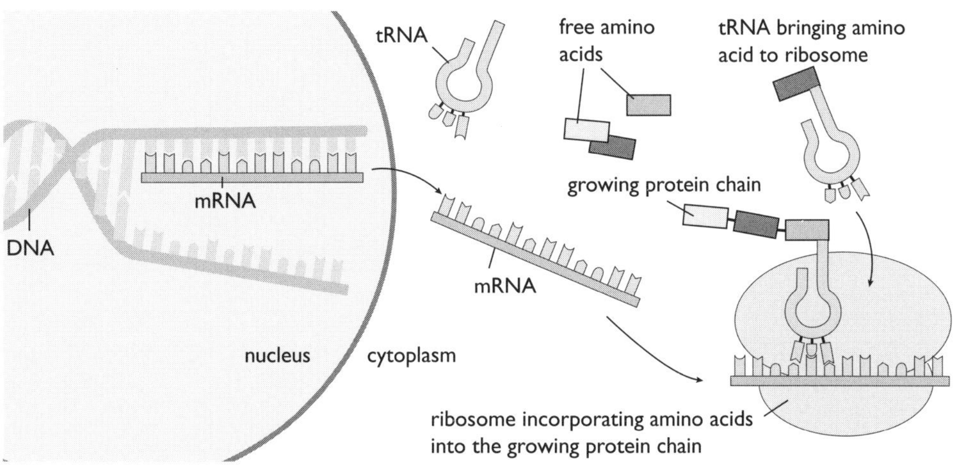

Figure 20.13 Relationships between the different types of RNA

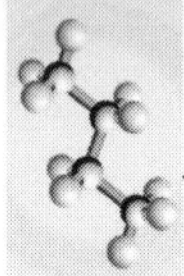

Protein synthesis

As we have seen, the ribosomes are the cellular 'factories' that synthesise proteins, and we have looked at the roles the different types of RNA play in this process. In order to understand protein synthesis fully we need to take a step or two back.

Both DNA and RNA each contain four bases. However, there are 20 amino acids used to make proteins. Therefore, more than one base is needed to code for a particular amino acid. If one pair of bases coded for a given amino acid there would only be 4^2 or 16 possible amino acids. As there are 20 amino acids, this means that three bases (a triplet) must be needed to code for every amino acid. This gives 4^3 or 64 possibilities.

Figure 20.14 shows how each of the 20 amino acids is coded, together with codes to stop the sequence — UGA, UAG and UAA — and one code AUG (Met) to start the sequence.

In Figure 20.14, the codes are read from the centre outwards.

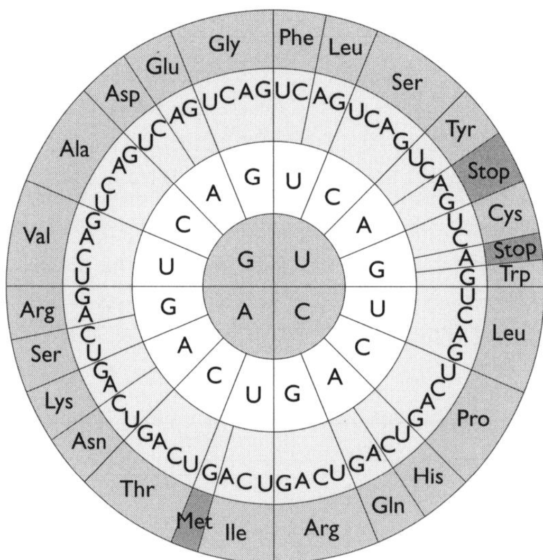

Figure 20.14 The genetic code

Try this yourself

(40) Which amino acid does each of the following base triplets code for? You can use the abbreviations in your answer.

(a) GAU

(b) CCA

(c) AAA

(41) Give one triplet code for each of the following amino acids:

(a) Ser (serine)

(b) Cys (cysteine)

(c) Tyr (tyrosine)

Amino acids cannot bind directly to mRNA. Transfer RNA molecules act as 'carriers', binding to a specific amino acid at one end of the tRNA molecule. At the other end of the tRNA molecule there is a triplet of bases (the anti-codon) that binds to the triplet of bases (the codon) on the mRNA. Each tRNA molecule interacts with the ribosome, enabling the amino acids to be joined to the growing chain by peptide bonds. This is summarised in Figure 20.15.

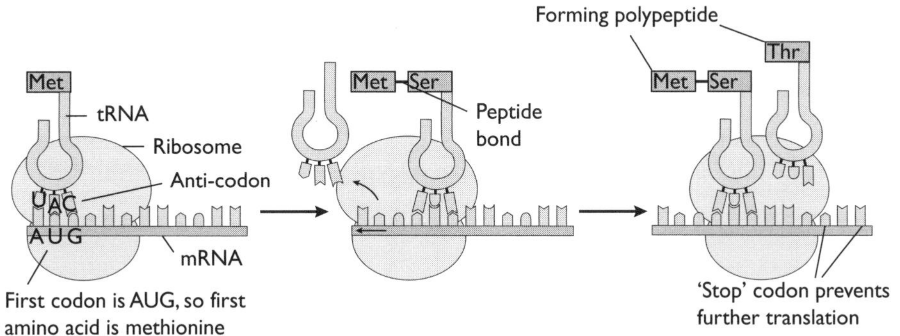

Figure 20.15 Process of synthesising a polypeptide in a ribosome

Mutations

Errors can occur during DNA replication and protein synthesis. However, it is relatively rare for these not to be corrected by mechanisms within the cell. Changes to the original DNA are known as **mutations**. They can occur naturally or can be caused by physical or chemical factors that damage DNA. These include ultraviolet light, ionising radiation, cigarette smoke and a range of other chemicals.

In many cases a single change to the base sequence will have little effect because most amino acids are coded for by a number of codons — for example, a change from GUA to GUG would still code for the amino acid valine (Val). However, if the mutation deleted a base from the sequence, this would produce an entirely different sequence of codons and hence a new amino acid sequence and a different protein. Mutation can result in genetic conditions such as sickle-cell anaemia and cystic fibrosis.

Sickle-cell anaemia

Sickle-cell anaemia is a condition that affects red blood cells making them crescent-shaped, rather than disc-shaped. The condition affects the haemoglobin and results in small blood vessels being blocked, preventing a good oxygen supply to some tissues. It is caused by a single mutation resulting in a change at the position of the sixth amino acid out of 146 in one of the protein chains in haemoglobin:

Normal chain	Val	His	Leu	Thr	Pro	Glu	Glu...
Sickle chain	Val	His	Leu	Thr	Pro	Val	Glu...

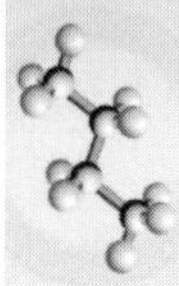

Cystic fibrosis

Cystic fibrosis is a relatively common genetic disorder. In some parts of the world it occurs in 1 in every 2000 live births. It has been estimated that around 1 in 22 Caucasians are carriers of the gene that causes it.

In cystic fibrosis, the normal fluid secretions in the lungs and gut are replaced by sticky mucus. It also affects the pancreas and restricts the supply of digestive enzymes. The sweat glands produce abnormally salty sweat. The reason for this condition is that a particular membrane protein responsible for the passage of chloride ions out of cells is either missing or faulty, thus preventing the release of chloride ions from cells.

It has been found that this is often the result of a particular triplet being deleted, resulting in a missing amino acid and hence a different sequence:

Normal chain	Ile	Ile	Phe	His	Lys...
Fibrosis chain	Ile	Ile	His	Lys...	

Things to remember

- There are two types of nucleic acid, DNA and RNA. They are formed from nucleotide monomers, each consisting of a sugar, a phosphate and a base.
- Both nucleic acids are condensation polymers with a sugar–phosphate backbone and the bases attached to the sugars forming side chains.
- DNA consists of a double stranded helix with the sugar–phosphate backbone on the outside and the bases pointing inwards towards each other. The two strands run in opposite directions and are held together by hydrogen bonds between the bases.
- There are four different bases in DNA, always paired in the same way: adenine–thymine and cytosine–guanine.
- RNA differs from DNA in that it:
  - contains the sugar ribose rather than deoxyribose
  - has a single strand
  - contains the base uracil rather than thymine
- DNA carries the genetic code for the production of proteins.
- Protein synthesis involves three types of RNA: messenger RNA (mRNA), transfer RNA (tRNA) and ribosomal RNA (rRNA).
- mRNA molecules contain a codon (triplet code) in which three bases in the RNA sequence code for one amino acid in the polypeptide chain.
- tRNA brings specific amino acids to the ribosome and binds to specific codons in the mRNA to allow the peptide bonds to form, extending the polypeptide chain.

Energy

You don't need to know a great deal about energy in biological systems for the A2 chemistry examinations. The syllabus focuses on adenosine triphosphate, ATP, which has an important role as a short-term energy source for metabolic reactions in living organisms.

The structure of ATP is shown in Figure 20.16. It is based on parts of molecules that you already know about:

- adenine — the base in DNA and RNA
- ribose — the sugar in RNA
- phosphate — the group found in the backbone of both DNA and RNA

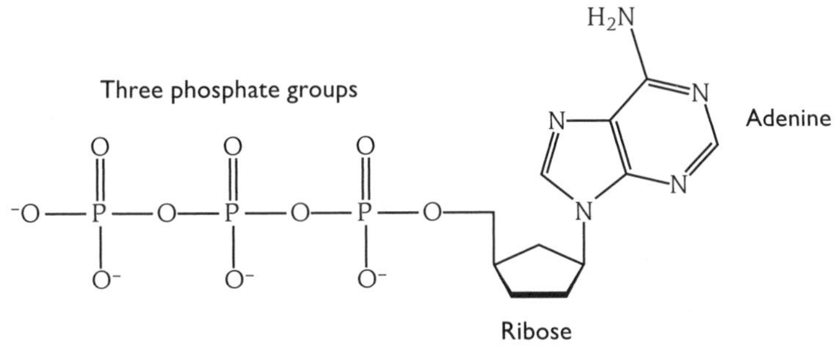

Figure 20.16 Structure of ATP

The hydrolysis of ATP is an exothermic reaction. The energy released is used by enzymes to drive catalysed reactions. The products of the reaction are ADP (adenosine diphosphate) and an inorganic phosphate ion, usually shown as P_i:

$$ATP + H_2O \rightarrow ADP + P_i \qquad \Delta H = -30\,kJ\,mol^{-1}$$

Metabolism uses ATP constantly, so its synthesis in the mitochondria of cells is a vital process. Each release of energy generates ADP, and ATP is re-formed from this. Figure 20.17 shows how this can happen.

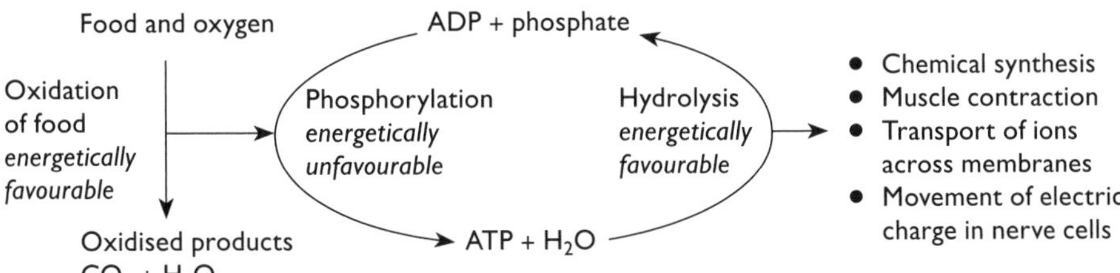

Figure 20.17 Regeneration of ATP

The oxidation of one molecule of glucose produces up to 38 molecules of ATP. The breakdown products of fats and proteins in our diet also feed into the respiratory pathway providing energy for cells via ATP. It is important to remember that while the hydrolysis of ATP is energetically favourable, the activation energy is normally high. It is only in the presence of the appropriate enzyme that the reaction occurs, enabling cellular processes to be controlled.

Things to remember

- All the chemical reactions involved in the function of cells and tissues are linked via metabolic pathways.
- ATP is a short-term energy source for metabolic reactions.

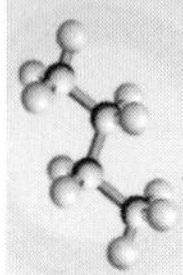

- Energy is released when ATP is hydrolysed to ADP and inorganic phosphate.
- The synthesis of ATP from ADP requires energy, which usually comes from the oxidation of food, and takes place in the mitochondria.

Metals in biological systems

Iron in haemoglobin

You are probably aware of the most well-known of the proteins that contain iron — haemoglobin. It is this protein complex that is responsible for carrying oxygen around the body in red blood cells. Haemoglobin consists of four protein chains, each bound to a non-protein haem group containing an iron(II), Fe^{2+}, ion. It is these Fe^{2+} ions that bind to the oxygen molecules. Each haem group can bind to one oxygen molecule, so each haemoglobin complex can carry four oxygen molecules (Figure 20.18). The arrangement is octahedral, and similar to the complexes with other transition metals (Chapter 11).

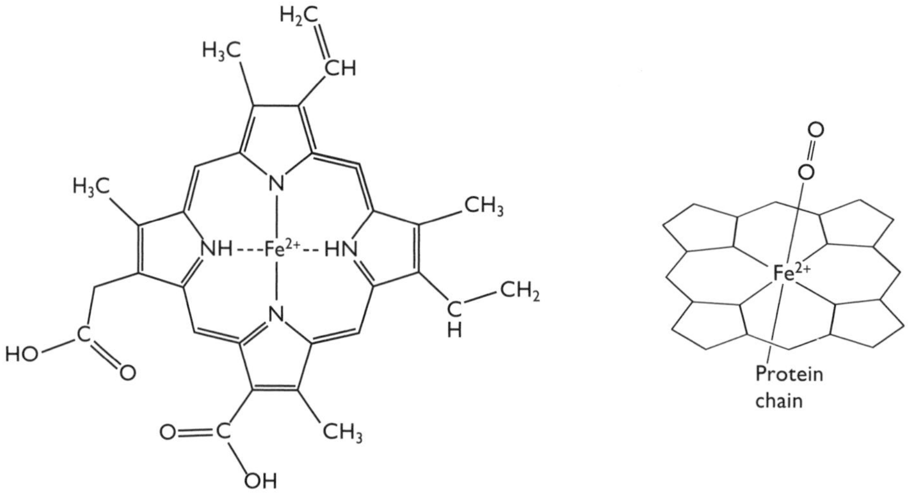

Figure 20.18 Haemoglobin

As with other transition metal complexes, the oxygen ligand can be replaced by other ligands. In the case of carbon monoxide this can prove fatal. Carbon monoxide binds to haemoglobin around 200 times more strongly than oxygen. This binding is effectively irreversible and results in haemoglobin being unable to exchange oxygen.

Sodium and potassium ions in membrane channels

The balance of ions within living cells is different from that of their surroundings. An example of this can be seen with sodium ions (Na^+) and potassium ions (K^+). The concentration of sodium ions within a cell is lower, and the concentration of potassium ions higher, than in the surrounding liquid. This is of particular importance in nerve cells. When a nerve is stimulated, sodium ions flood into the cell. When the 'signal' has passed it is vital to restore the original balance ready for another nerve impulse.

The energy for these changes comes from the hydrolysis of ATP. This change is accomplished by the use of an ion-transporting enzyme known as 'Na^+, K^+-ATPase'. This is not a very helpful name and so it is often referred to as the 'sodium–potassium pump'. These enzyme molecules are found bridging cell membranes and control the flow of ions in and out of the cells.

Zinc as an enzyme cofactor

The enzyme carbonic anhydrase is one of the most efficient enzymes in the human body. It is responsible for the removal of carbon dioxide from the blood within the red blood cells by producing hydrogencarbonate ions. Cofactors act by enabling the substrate to react at the active site (Figure 20.19).

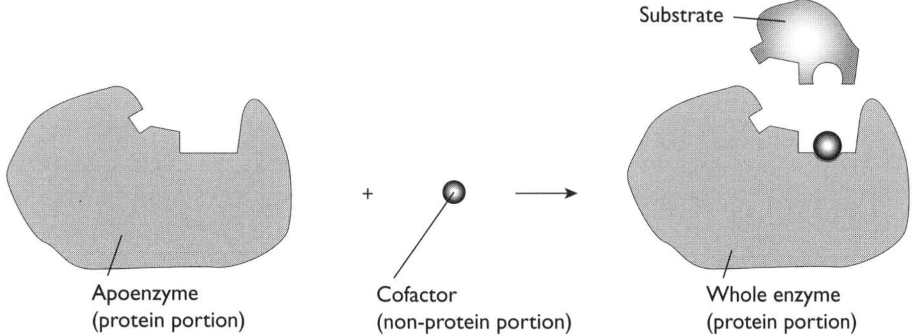

Figure 20.19 Action of a cofactor

In the case of carbonic anhydrase, the presence of a high charge density from the Zn^{2+} ions at the active site assists in the breakdown of a water molecule into H^+ and OH^- ions. The OH^- ions then attack the carbon dioxide forming hydrogencarbonate ions:

$$CO_2 + OH^- \rightarrow HCO_3^-$$

Once the hydrogencarbonate is released, another water molecule binds to the zinc ion allowing the cycle to begin again.

Toxic metals

A number of metals, particularly 'heavy metals' (lead, mercury, cadmium and arsenic) can have serious effects on health even though they may be present in only small amounts in the environment. One of the reasons for this is that they can accumulate in the food chain and, therefore, may be ingested at significantly higher levels than those found in the environment.

For example, mercury can disrupt disulfide bridges and other 'heavy metals' have been shown to disrupt van der Waals forces. Lead is known to cause mental health problems in children, and mercury can cause major failure of the nervous system. The effects of mercury are sometimes referred to as 'Minamata disease' after the Japanese city in which over 1700 people died after methyl mercury was released in

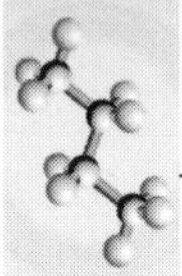

waste water. The mercury accumulated in the food chain via shellfish and fish which were then consumed by the local population.

Things to remember

- Metals have key roles in the functioning of biological systems.
- Iron is important in proteins that use the haem group; the balance of sodium and potassium ions is vital for maintaining cell structure and the transmission of nerve impulses; zinc is important as a cofactor in the key enzyme carbonic anhydrase.
- Metals such as lead and mercury are toxic due to the effects they have on the tertiary structure of proteins, particularly van der Waals forces and disulfide bridges.
- The effects of heavy metals are made worse because they can accumulate in the food chain even if within the environment their concentrations are small.

Applications of analytical chemistry

The examination of this part of the syllabus generally focuses on the use of the different analytical techniques, particularly in terms of analysing data to provide structural information about compounds. Your knowledge of core chemistry will help you to do this.

Methods of detection and analysis

There are five main areas of analysis that you need to be familiar with:

- electrophoresis and DNA fingerprinting
- proton-NMR spectroscopy
- X-ray crystallography
- chromatography
- mass spectrometry

In each case you need to be able to explain how the technique works and what information it provides. You may be expected to interpret data from one or more of these techniques to solve a problem.

Electrophoresis and DNA fingerprinting

We saw in Chapter 18 that amino acids form zwitterions in aqueous solution, and that the exact nature of a zwitterion depends on the pH of the solution:

$H_3N^+—CH_2—CO_2^-$

zwitterion

Low pH: $H_3N^+—CH_2—CO_2^- + H_3O^+ \rightarrow H_3N^+—CH_2—CO_2H + H_2O$

High pH: $H_3N^+—CH_2—CO_2^- + OH^- \rightarrow H_2N—CH_2—CO_2^- + H_2O$

This effect can be seen when an electric potential is applied to a mixture of the amino acids glycine, lysine and glutamic acid at pH 7 using the set-up shown in Figure 20.20.

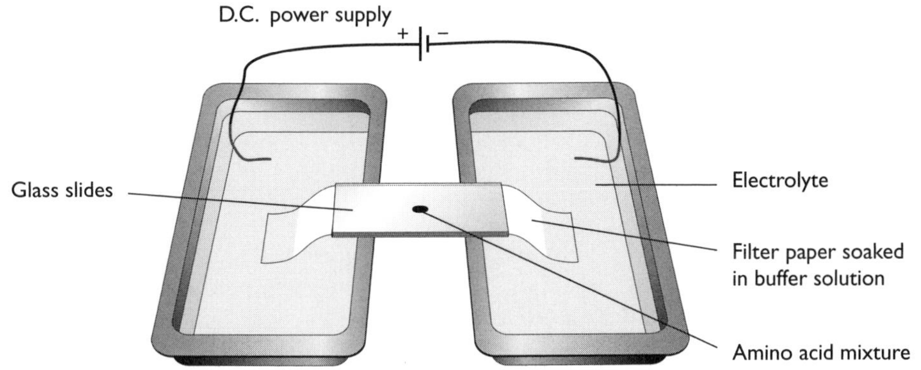

Figure 20.20 Apparatus for electrophoresis

If the power supply is connected and the apparatus left for a period of time the amino acid mixture is separated as shown in Figure 20.21.

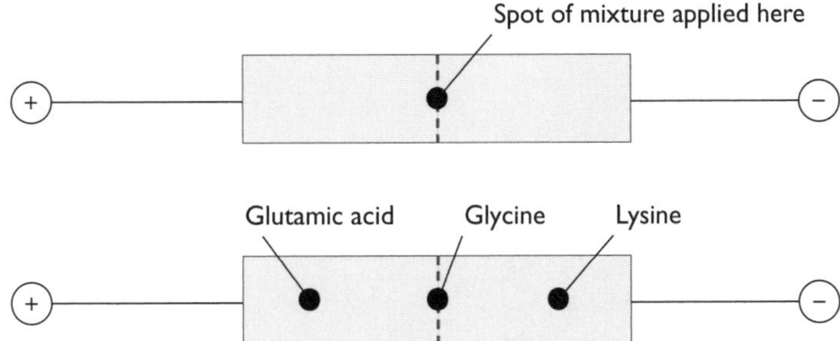

Figure 20.21 Results of the electrophoresis of a mixture of three amino acids

We can explain how different amino acids are affected by looking at the structures of glycine, lysine and glutamic acid at pH 7:

glycine	$H_3N^+CH_2CO_2^-$
lysine	$H_3N^+CH(CH_2CH_2CH_2CH_2NH_3^+)CO_2^-$
glutamic acid	$H_3N^+CH(CH_2CH_2CO_2^-)CO_2^-$

Note that at pH 7:

- lysine carries an extra positive charge, and hence moves towards the negative electrode
- glutamic acid carries an extra negative charge and moves towards the positive electrode
- glycine carries one of each type of charge so it is attracted equally to both electrodes and does not move

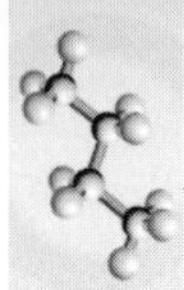

The same process can be used to separate protein fragments (peptides).

The velocity of the amino acids is related directly to the voltage applied across the plate. Other factors can affect the relative velocities of samples:

- Smaller molecules move faster than larger molecules carrying the same charge.
- A molecule with large side chains moves more slowly than a straight-chain molecule with the same charge and M_r.
- The pH of the buffer influences the extent of ionisation, and hence movement and direction.

DNA fingerprinting

This has become a 'high-profile' analytical technique due largely to its use in forensic science to screen crime suspects. It uses the same technique as electrophoresis, but rather than relying on filter paper as the medium, a gel is used. In recent years the technique has been refined so that only a tiny amount of DNA is needed. This can be extracted from material such as blood, hair, cheek cells, semen or skin.

The DNA is treated with enzymes to break it into fragments that can be analysed by electrophoresis. If the amount of DNA is very small, a technique called the polymerase chain reaction (PCR) can be used to produce more copies of the DNA. It is possible to start the analysis with as little as 0.2 nanograms (2×10^{-10} grams) of DNA.

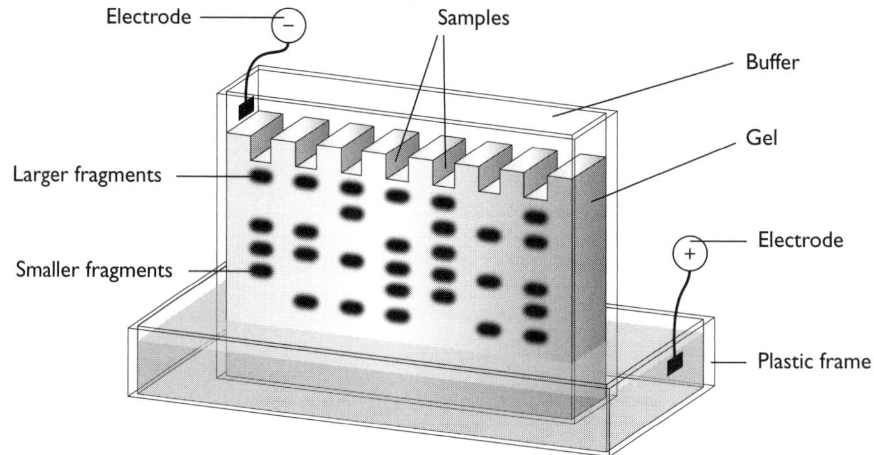

Figure 20.22 Gel electrophoresis

Figure 20.22 shows the apparatus for carrying out gel electrophoresis. Samples of the DNA fragments are put into small 'wells' in the gel near the cathode. The phosphate groups on the DNA fragments are negatively charged and are attracted to the anode, with smaller fragments moving faster than larger ones. The results can be stained to help them show up or, if the DNA fragments are treated with radioactive phosphorus, a photographic print can be obtained. A typical example is shown in Figure 20.23.

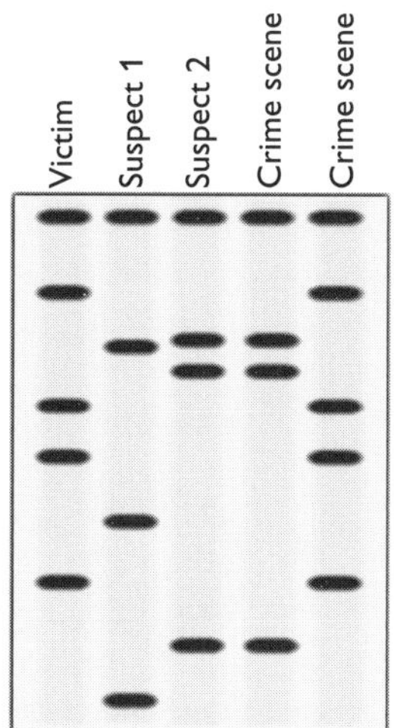

Figure 20.23 A simple example to show the results of gel electrophoresis

In this simplified example it is relatively straightforward to pick out the DNA of both the victim and suspect 2 from the DNA samples collected at the crime scene.

Although the use of DNA in crime detection gets a great deal of publicity, it has a number of other important uses:

- establishing the father of a child in a paternity case
- making links within a family from samples obtained from both living and deceased relatives
- establishing links between archaeological samples of biological origin, such as animal skins
- medical applications where the presence of a particular polypeptide or protein can be an early indication of a problem without symptoms, for instance in newborn babies

Things to remember

- Amino acids may be positively charged, negatively charged or neutral. This means they can be separated on buffered filter paper when an electric potential is applied.
- The speed and direction of movement in electrophoresis depends on the size of the amino acid or peptide, the size of any side chains present and the charge carried at the pH of the buffer used.
- DNA is broken into fragments using enzymes called **restriction enzymes**.
- The amount of DNA can be increased using the polymerase chain reaction (PCR).
- The phosphate groups in DNA fragments are negatively charged enabling them to move when an electric potential is applied.
- DNA fingerprinting has a range of applications including providing forensic evidence, establishing genetic links between individuals, archaeology involving specimens of biological origin and early diagnosis of genetic diseases.

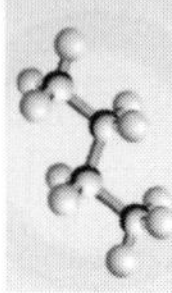

Proton-NMR spectroscopy

The nucleus of a hydrogen atom spins about an axis. Since the nucleus is positively charged, this spinning produces a magnetic field giving each nucleus a magnetic moment. If an external magnetic field is applied, the nuclei align their magnetic moments parallel to the applied field (Figure 20.24 (a)). Another possible state exists where the magnetic moments are aligned against the applied field, but this requires extra energy (Figure 20.24(b)).

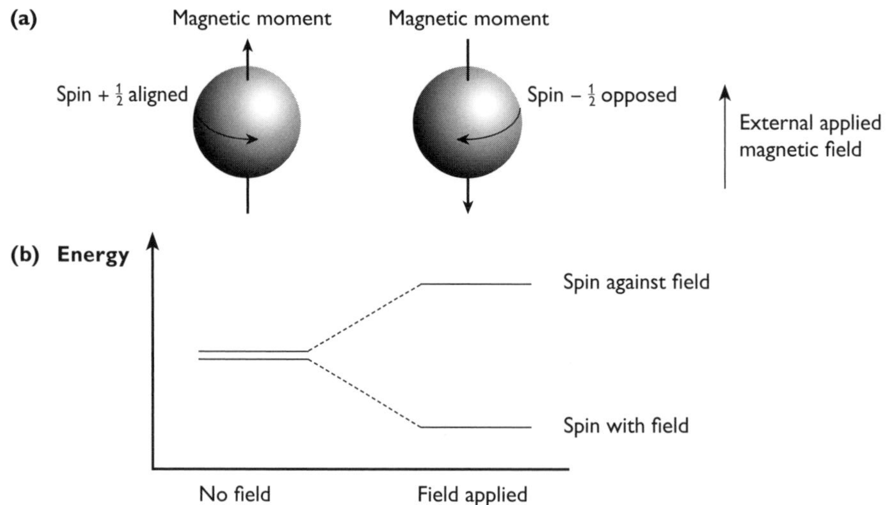

Figure 20.24 Magnetic moments

Just looking at hydrogen atoms is not particularly useful, but hydrogen atoms in organic molecules are influenced by the atoms adjacent to them. Not all protons absorb energy at the same frequency when 'flipping' their magnetic moment to oppose the applied magnetic field. The external magnetic field applied is modified by the different chemical environments in the molecule. For example, the presence of an electronegative atom causes the bonding electrons to be attracted towards it, leaving any protons less shielded from the external field. This causes the energy change for 'flipping' the proton magnetic moment to be at a higher frequency.

To obtain the NMR spectrum of a compound, it is first dissolved in a solvent that does not absorb in the proton-NMR region and then placed in a magnetic field. Energy is supplied at radio frequency by scanning over a range. Carrying out this procedure on ethanol at low resolution results in three distinct absorptions due to the three protons in different environments (Figure 20.25). A small amount of tetramethylsilane (TMS) is added to give a marker at 0.05.

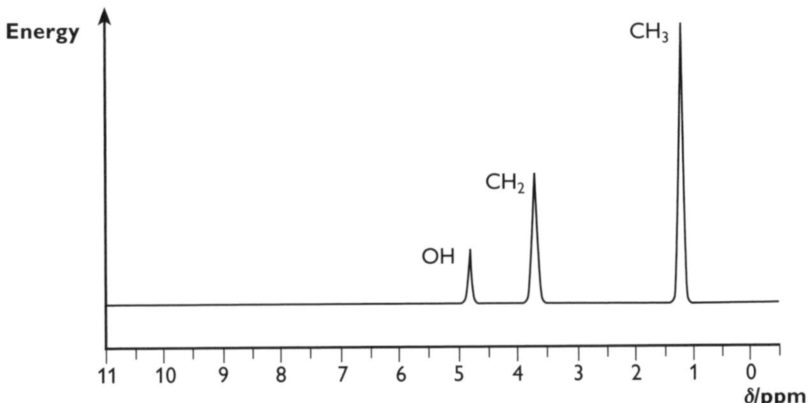

Figure 20.25 Low resolution NMR spectrum of ethanol

Note that the peaks are not only in different positions, but that the areas under the peaks are different. The proton attached to the electronegative oxygen atom is less shielded and so absorbs at higher frequency, which in an NMR spectrum is to the *left*. The scale, given the symbol δ, is measured in parts per million. The protons affected least by the oxygen are the three protons in the methyl group at the other end of the molecule and are shown by the peak furthest to the right. This is also the peak of greatest area, since it is produced by three protons in an identical chemical environment.

The spectrum for ethanol at high resolution (Figure 20.26) is a little different. Two of the peaks are now split into smaller peaks. This occurs because protons are not only influenced by their chemical environment, but also by the magnetic moments of adjacent protons. This is called **spin–spin splitting**.

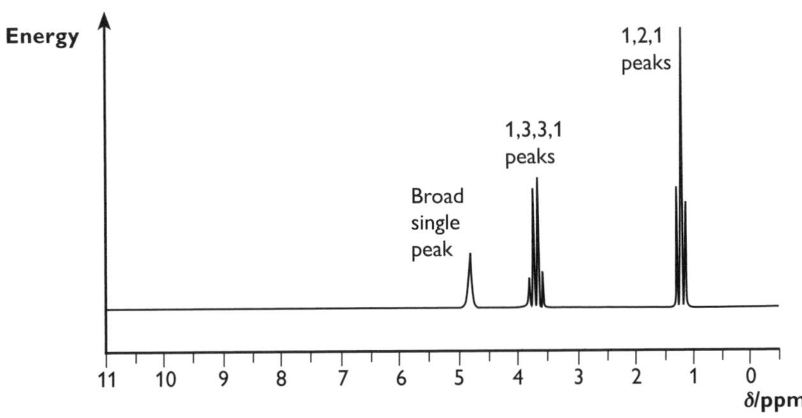

Figure 20.26 High resolution NMR spectrum of ethanol

Consider the protons in the methyl, $-CH_3$, group. They are influenced by the protons on the adjacent methylene, $-CH_2-$, group. These two protons could both align with the applied field, both against the applied field, or one with and one against the

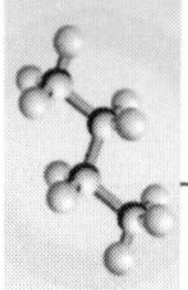

applied field — and there are two ways of doing this! This results in the methyl absorption being split into three peaks in the ratio 1:2:1:

↑↑ ↑↓ ↓↓

↓↑

The absorption due to the methylene protons is split into four peaks in the ratio 1:3:3:1 by the adjacent methyl protons. You might expect that the –OH proton peak would also be split by the methylene protons. However, this does not happen because these protons exchange rapidly with other –OH protons present. A useful test for –OH protons is to add D_2O (deuterium oxide) to the sample. Deuterium nuclei do not absorb in the same range as normal protons, so if an –OH group is present, its peak disappears when D_2O is added.

Table 20.3 shows possible splitting patterns produced by protons adjacent to the one being considered.

Table 20.3

Number of protons adjacent to given proton	Number of lines in multiplet	Relative intensities of lines
1	2	1:1
2	3	1:2:1
3	4	1:3:3:1
4	5	1:4:6:4:1

Things to remember

- The nuclei of hydrogen atoms spin about an axis producing a magnetic moment.
- In an external magnetic field there are two energy states — aligned with and against the external field. If energy is supplied in the radio frequency range the protons can 'flip' to the higher energy state (aligned against the external field).
- The field experienced by a given proton is modified by fields produced by different chemical environments (chemical shift). Protons adjacent to electronegative atoms have absorptions moved to higher frequency (δ).
- The field is further modified by the magnetic moments of adjacent protons (spin-spin splitting). These splitting patterns are characteristic of alkyl groups e.g. –CH gives 1:1, $-CH_2$ gives 1:2:1 and $-CH_3$ gives 1:3:3:1.
- Protons in –OH groups usually show only one peak because they exchange with other –OH protons present. Adding D_2O makes this peak disappear.

X-ray crystallography

X-rays are very short wavelength electromagnetic radiation. The wavelengths are similar to the interatomic distances in solids. If a beam of monochromatic X-rays (i.e. rays with a single wavelength) is directed at a crystal, some of the X-rays are diffracted by the planes of atoms in the crystal. This produces a pattern of intensities that can be interpreted to give electron density maps of the crystal. The most concentrated electron density is around the largest atoms (the ones with the most electrons).

This technique allows us to measure bond lengths and bond angles of molecules, but does *not* show up hydrogen atoms since they only possess one electron and are swamped by other larger atoms.

Things to remember

- X-rays are diffracted by planes of atoms in a crystal because they have similar wavelengths to the separation of the atoms in molecules.
- The intensity of the interaction depends on the electron density around a given atom. Due to this, hydrogen does not show up.
- Using X-ray crystallography, bond lengths and bond angles in molecules can be measured.

Chromatography

Partition

In order to understand how chromatography works, you first need to understand what is meant by **partition**. You are familiar with water as a solvent, but many other liquids also act as solvents. In general, substances dissolve when the energy of the solute–solvent system is lower when the solute is dissolved than when the solute is not dissolved. This is usually the case if the interactions between the solute and the solvent molecules are similar to those between the solvent molecules themselves.

Molecules can attract one another in a variety of ways:

- ionic attractions
- ion–dipole attractions
- hydrogen bonding
- van der Waals forces

As a rule:

- polar solvents are more likely to dissolve ions, substances that form hydrogen bonds and molecules with dipoles
- non-polar solvents dissolve solutes whose molecules are attracted to each other by only van der Waals forces

Iodine is a molecular solid and is unlikely to be as soluble in water, a polar solvent, as it is in hexane, a non-polar solvent. If some iodine crystals are shaken with a mixture of water and hexane until no further change occurs, the iodine distributes itself between the two solvents according to its solubility in each. On measuring the amount of iodine dissolved in each solvent, we find that the ratio of the concentrations is constant, no matter how much iodine is used. This constant is known as the **partition coefficient, K_{pc}**.

$$K_{pc} = \frac{[I_{2(hexane)}]}{[I_{2(water)}]}$$

Here, K_{pc} is an equilibrium constant for the dissolving of iodine in the two solvents. Partition coefficients generally have no units.

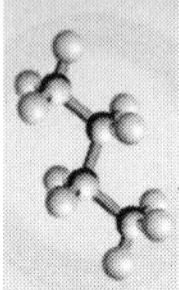

Paper chromatography

You may wonder what this has to do with chromatography, but paper chromatography works because the solutes partition themselves between water held on the fibres of the paper and the solvent that is being used. This means that solutes which are more soluble in the solvent are carried further.

Components in a mixture can be identified by running reference samples alongside the mixture or by calculating the **retardation factor, R_f**. This is the ratio of the distance travelled by the spot to the distance travelled by the solvent (Figure 20.27).

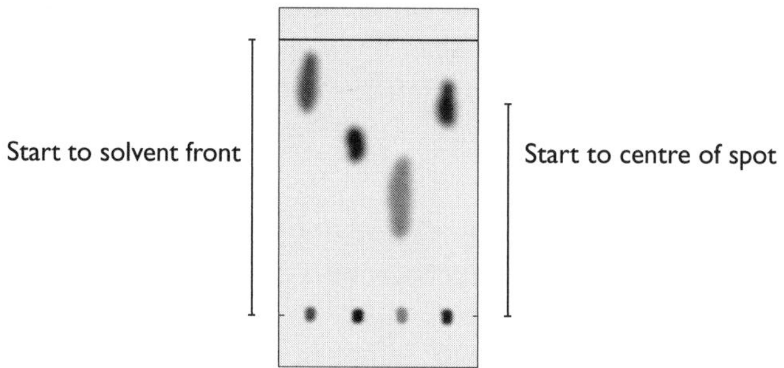

Figure 20.27 Data needed to calculate R_f

You are probably familiar with using paper chromatography to separate the dyes in different inks, and have perhaps obtained a chromatogram like the one shown in Figure 20.28.

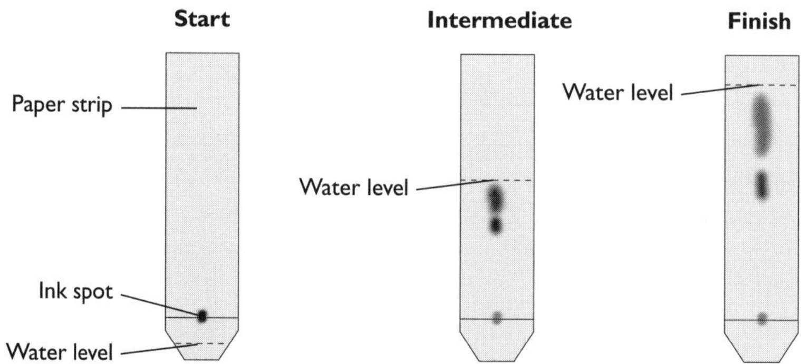

Figure 20.28 Paper chromatography of black ink

Some mixtures may not give coloured spots. For example, if we try to separate a mixture of amino acids by chromatography the individual amino acids are invisible. To reveal the separation we have to use a 'locating agent', in this case ninhydrin, which turns the spots purple-pink (Figure 20.29).

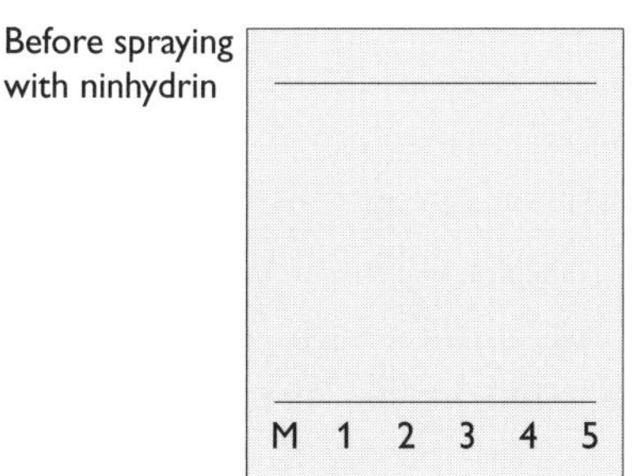

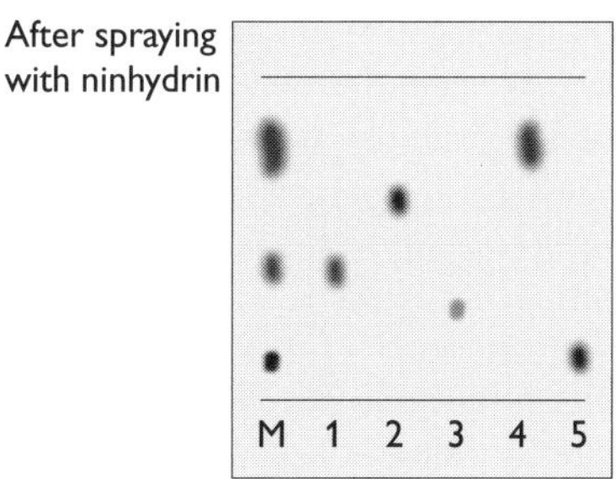

Figure 20.29 Use of a locating agent in paper chromatography

In Figure 20.29, reference samples of amino acids (1–5) have been run alongside the mixture so there is no need to calculate R_f values.

Thin-layer chromatography

This method relies on the adsorption of the solute onto the particles of the thin layer, rather than partition between water trapped on the cellulose and the solvent. Figure 20.30 highlights the differences.

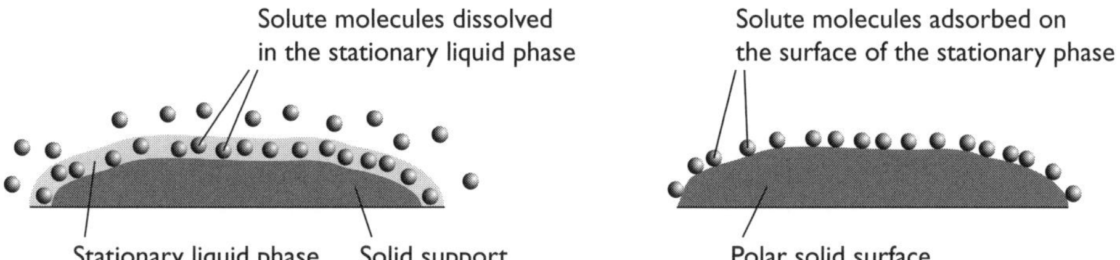

Figure 20.30 Separation by (a) partition and (b) adsorption

Gas-liquid chromatography (GLC)

In GLC, a gas is used as the mobile phase and a non-volatile liquid held on small inert particles is the stationary phase. These particles are packed into a long column a few millimetres in diameter and up to 3 metres long. The column is coiled and mounted in an oven whose temperature can be controlled. This technique works for samples that are gases or which have significant vapour pressure at the temperature of the oven.

The sample is injected into the column and vaporised if necessary. The separation takes place in the column and the individual components emerge from the end of the column and are detected. Sometimes the GLC column is connected to a mass spectrometer (see next section). Different components take different times to flow through the columns. These are known as **retention times** and depend on the following factors:

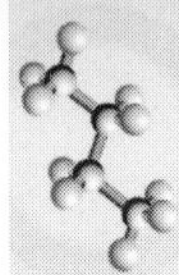

- the dimensions (length and diameter) of the column
- the temperature of the column
- the flow rate of the carrier gas
- the volatility of the sample
- the interactions between the components of the sample with both the mobile and stationary phases

The retention times may help to identify individual components. By measuring the area under each peak, the proportions of the different components in the sample can be determined.

High performance liquid chromatography (HPLC)

This technique is similar to GLC. The main differences are:

- the mobile phase is a liquid rather than a gas
- the liquid is forced through the column at pressures up to 400 atm
- the columns are much shorter (10–30 cm)
- the components are usually detected by measuring the absorbance of ultraviolet radiation through a cell at the end of the column

As with GLC, the retention times of different substances depend on a number of factors:

- the temperature of the column
- the pressure used (because that affects the flow rate of the solvent)
- the nature of the stationary phase (not only what material it is made of, but also particle size)
- the exact composition of the solvent

Things to remember

- When a solute is shaken with two immiscible solvents, it will dissolve in both. The ratio of the concentration of the solute in each is called the 'partition coefficient'.
- Chromatography is based on the ability of a solute to dissolve preferentially in one of two solvents.
- Components in a chromatogram can be identified by their retardation factor, R_f — the ratio of the distance travelled by the substance compared with the distance travelled by the solvent. Reference samples can also be used to aid identification.
- For colourless samples, such as amino acids, a locating agent is used to make the spots visible.
- In GLC, compounds can be identified by their retention times. Measuring the area under the peaks enables the proportions of each substance in a mixture to be determined.

Mass spectrometry

The principles on which mass spectrometry is based are simple, and modern mass spectrometers (Figure 20.31) are highly accurate having an accuracy greater than 1 part in 100 000.

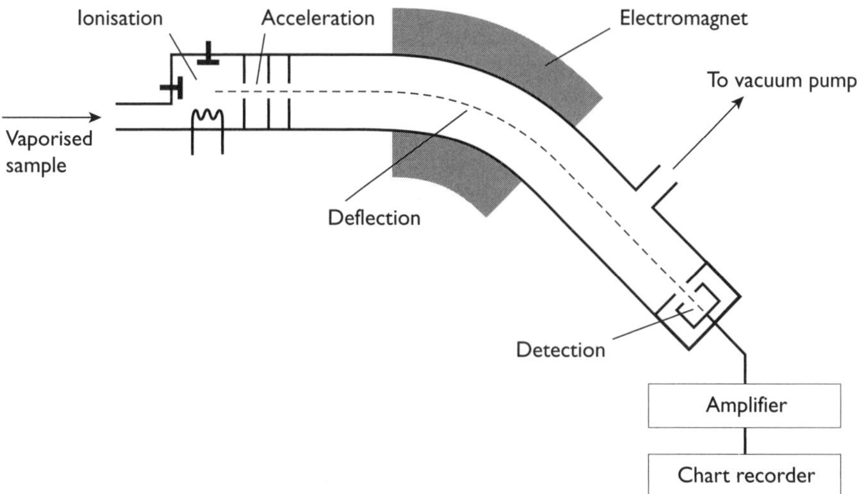

Figure 20.31 A mass spectrometer

In the mass spectrometer the following processes occur:

- If it is not already a gas, the compound is vaporised.
- It is bombarded with electrons, which knock other electrons off some of the atoms or molecules forming positive ions: $M + e^- \rightarrow M^+ + 2e^-$
- Gaseous ions are accelerated in an electric field and then pass through an electrostatic analyser forming a tightly focused beam.
- The ions pass between the poles of an electromagnet, which deflects the beam. Lighter ions are deflected more than heavier ions.
- The strength of the magnetic field is adjusted to allow ions to pass through a slit and be detected.

We saw in Chapter 1 how mass spectrometry can be used to determine accurate atomic masses. It can also be used to give useful information about organic molecules. Mass spectrometry can give three different types of information:

- Measurement of the relative heights of the M and M + 1 peaks allows us to determine the number of carbon atoms in a molecule. Examination of the M and M + 2 peaks can identify the presence of chlorine or bromine.
- Measurement of the accurate mass of the molecular ion (the M peak) enables us to determine the molecular formula.
- Identification of fragment ions produced in the mass spectrometer may allow the piecing together of the structure of the parent molecule.

Use of the M + 1 peak

Naturally-occurring carbon consists almost entirely of ^{12}C and ^{13}C in the ratio 98.9% ^{12}C to 1.1% ^{13}C. We can use this information, together with the abundance of the M and M + 1 peaks, to calculate the number of carbon atoms, n, in a molecule.

$$n = \frac{100\ (A_{M+1})}{1.1\ (A_M)}$$

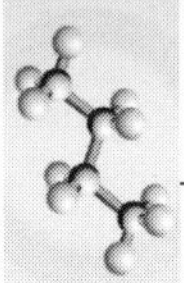

where n is the number of carbon atoms and A_M and A_{M+1} are the abundances of the M and M + 1 peaks.

Use of the M and M + 2 peaks

The halogens chlorine and bromine occur naturally as mixtures of two predominant isotopes. This is shown in Table 20.4.

Table 20.4

Element	**Isotope**	**Relative abundance**	**Approximate. ratio**
Chlorine	^{35}Cl	75.8%	3:1
	^{37}Cl	22.4%	
Bromine	^{79}Br	50.5%	1:1
	^{81}Br	49.5%	

You can see that if the ratio of the abundance (height) of the M and M + 2 peaks is 3:1 it indicates the presence of a chlorine atom in the molecule. If the ratio is 1:1 it indicates the presence of a bromine atom. If more than one atom of halogen is present there would also be an M + 4 peak and we would be able to determine which halogen atoms are present from the ratios.

Using accurate molecular masses

With high resolution mass spectrometers we can measure m/e ratios to five significant figures (at least 1 part in 100000). This means that it is not only possible to accurately measure the M_r of a compound, but also to determine its molecular formula.

The two compounds below have the same M_r to the nearest whole number:

Pent-1-ene	C_5H_{10}	$CH_3CH_2CH_2CH{=}CH_2$
But-1-ene-3-one	C_4H_6O	$CH_2{=}CHOCH_3$

Using the accurate relative atomic masses for hydrogen, carbon and oxygen — H = 1.0078, C = 12.000, O = 15.995 — we can determine the accurate relative molecular masses of the two compounds.

$$C_5H_{10} = 5 \times 12.000 + 10 \times 1.0078 = 70.078$$

$$C_4H_6O = 4 \times 12.000 + 6 \times 1.0078 + 15.995 = 70.0418$$

Using fragmentation patterns

The electron beam that produces ionisation in the mass spectrometer can also break bonds, producing fragments of the parent molecule. Some of these fragments will have a positive charge and produce further peaks in the mass spectrum. The fragmentation pattern can help to distinguish between structural isomers because they form different fragments. The two mass spectra shown in Figure 20.32 were obtained from propan-1-ol (Figure 20.32(a)) and propan-2-ol (Figure 20.32(b))

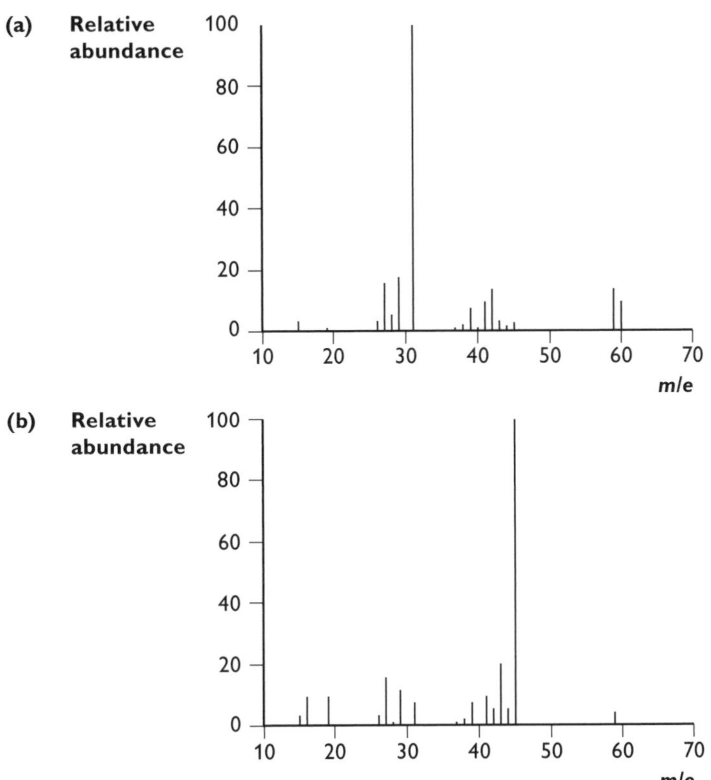

Figure 20.32 Mass spectrum of (a) propan-1-ol and (b) propan-2-ol

The mass spectra have very different fragmentations. In propan-1-ol the largest peak is at *m/e* 31, whereas in propan-2-ol the largest peak is at *m/e* 45. What are the fragments that cause these two peaks and how are they formed from the parent molecules? Let's start by looking at the two isomers:

Propan-1-ol

Propan-2-ol

Look at the spectrum of propan-1-ol. A break in the middle of the molecule (where the arrow is) produces either $CH_3CH_2^+$ and $\bullet CH_2OH$ or $^+CH_2OH$ and $CH_3CH_2\bullet$. The uncharged species do not appear in the mass spectrum; the charged species have peaks at *m/e* values of 29 and 31.

In propan-2-ol, fragmentation between the first two carbon atoms produces either CH_3^+ (*m/e* = 15) or $^+CH(OH)CH_3$ (*m/e* = 45) as the charged species.

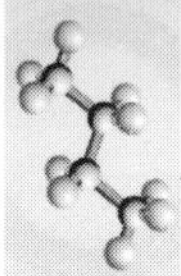

The fragments in each case can be found in the relevant spectrum in Figure 20.32. For the examination you do not need to be able to explain *why* particular fragments form, just to be able to identify *what* their formula might be.

Things to remember

- In a mass spectrometer, compounds are vaporised, turned into positive ions by bombardment with electrons, accelerated by an electric field and separated by mass using a magnetic field.
- In organic compounds, the ratio of the abundances of the M and M + 1 peaks can be used to determine the number of carbon atoms present, since 1.1% of naturally-occurring carbon is ^{13}C. The expression used is:

$$n = \frac{100\ (A_{M+1})}{1.1\ (A_M)}$$

- If the mass spectrum contains an M + 2 peak, either chlorine or bromine is present in the molecule. If the ratio is 3:1 chlorine is present; 1:1 indicates that bromine is present.
- A highly accurate mass spectrum can identify the molecular formula of a compound.
- The fragmentation pattern of a molecule can help determine the structure of a compound.

Applications in chemistry and society

In this section we have looked at a range of modern analytical techniques and the ways in which they may be used. The syllabus refers to another area in which analytical techniques play an important role — that of monitoring the environment. The two examples below illustrate this.

Monitoring PCBs in the atmosphere

PCBs (polychlorinated biphenyls) are a group of man-made chemicals that have been used in a variety of manufactured goods since the late 1920s. They can persist in the environment and accumulate in the food chain. PCBs can be found around waste sites, particularly incinerators, where they can be converted into similarly dangerous dioxins. Many countries have banned or severely restricted the production of PCBs.

Humans may be exposed to PCBs by consuming contaminated food, drinking contaminated water or breathing contaminated air. Mothers exposed to PCBs may transmit them to their unborn child or to breast-feeding infants. Effects of PCBs include an increased risk of some forms of cancer, reduced fertility in men and some neurological effects.

PCBs can be monitored by taking air samples remotely, separating the chemicals found using GLC and then analysing them by mass spectrometry. They can also be monitored in water and in soil samples.

Measuring isotope ratios in ice cores

Oxygen is one of the key markers when studying past climates. Natural oxygen exists as two main isotopes — ^{16}O and a tiny amount of ^{18}O, which contains two extra neutrons. These two isotopes can be detected and the ratio measured using mass spectrometry.

The ratio of these isotopes of oxygen in water changes with the climate. By determining the ratio of ^{16}O and ^{18}O oxygen in marine sediments, ice cores and fossils and comparing this with a universally accepted standard, scientists can study climate changes that have occurred in the past. The standard used for comparison is based on the ratio of oxygen isotopes in ocean water at a depth of between 200 and 500 metres.

Evaporation and condensation are the two processes that most influence the ratio of ^{16}O and ^{18}O oxygen in the oceans. Water molecules containing ^{16}O evaporate slightly more readily than water molecules containing ^{18}O. Similarly, water vapour molecules containing ^{18}O condense more readily.

During ice ages, cooler temperatures extend toward the equator, so the water vapour containing ^{18}O rains out of the atmosphere at lower latitudes than it does under milder conditions. The water vapour containing ^{16}O moves toward the poles, eventually condenses, and falls onto the ice sheets where it stays. The water remaining in the ocean develops an increasingly higher concentration of ^{18}O compared with the universal standard, and the ice develops a higher concentration of ^{16}O. Thus, high concentrations of ^{18}O in the ocean tell scientists that ^{16}O was trapped in the ice sheets. The exact oxygen ratios can indicate how much ice covered the Earth. This work can also provide pointers to any current change in climate by examining the isotopic ratio in recent ice cores.

Design and materials

You may be asked to use your knowledge of core chemistry in discussing new materials and chemical techniques. This section focuses on modern developments in the areas of drug design and delivery, new materials, manipulation of materials on a molecular scale (nanotechnology), environmental problems and extending the useful life of both materials and energy sources.

Medicinal chemistry and drug delivery

Drug design

Huge quantities of money are invested each year by pharmaceutical companies on research into new drugs and treatments for major diseases. The cost is often the result of promising candidates that prove to be unsatisfactory at the clinical trial stage after many months or years of development.

In modern research, careful preparation and analysis is carried out so that the range of possible molecules to be tested is limited and that unsuitable ones are screened out at an early stage. However, sometimes only tiny structural differences can make a huge difference to the effect of the drug. One such example is the drug thalidomide,

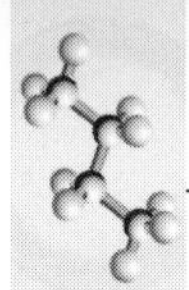

which was produced by researchers seeking an alternative sedative to Valium® for women during pregnancy. Like many organic compounds, thalidomide exists in the form of two optical isomers:

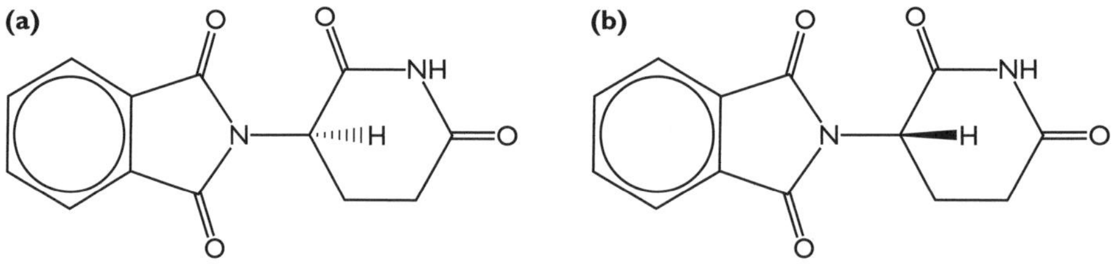

(a) (b)

Figure 20.33(a) R-thalidomide and (b) S-thalidomide

The difference in the two forms is slight, but it was sufficient for the S-form to cause serious birth defects, whereas the R-form had the desired properties of being a sedative and anti-nausea drug.

Drug molecules act by binding to receptors, which are often enzymes. You will remember how only particular substrate molecules fit into the active site and bind to it. Molecules that do not have the precise shape and binding sites are either inactive or have different effects. A simple example of this is the compound carvone. This is an optically active compound that binds to sites in the olfactory system (the region responsible for detecting smells). The two isomers (Figure 20.34) have quite different smells:

(a) (R)-(–)-carvone from spearmint oil

(b) (R)-(+)-carvone from caraway seed oil

Figure 20.34 Isomers of carvone

In modern drug research, chemists use computer models to see how a particular drug might interact with the active site on the receptor. They also have access to online databases to check on interactions with other enzymes in the body. These techniques, while not cheap in themselves, have refined drug design and research making new powerful treatments for major diseases available in a shorter time.

The drug Taxol® is based on a natural substance found in yew trees. It is used to treat some forms of cancer. If you study its structure (Figure 20.35) you will recognise that it contains a number of amide and ester linkages, all of which can be hydrolysed in acid. Based on this it would not be sensible to give the drug orally without protecting it in some way.

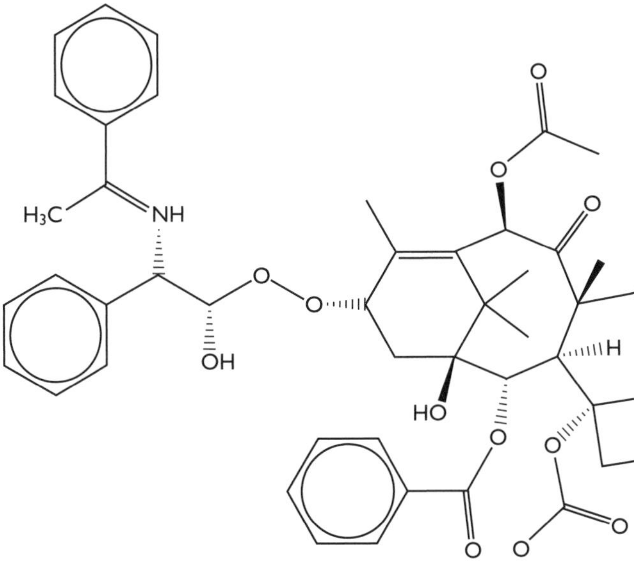

Figure 20.35 The structure of Taxol®

One successful method of delivery that has become widespread is to use artificial liposomes (Figure 20.36). These are tiny spheres made from one or more phospholipid bilayers. They contain an aqueous core that houses the drug.

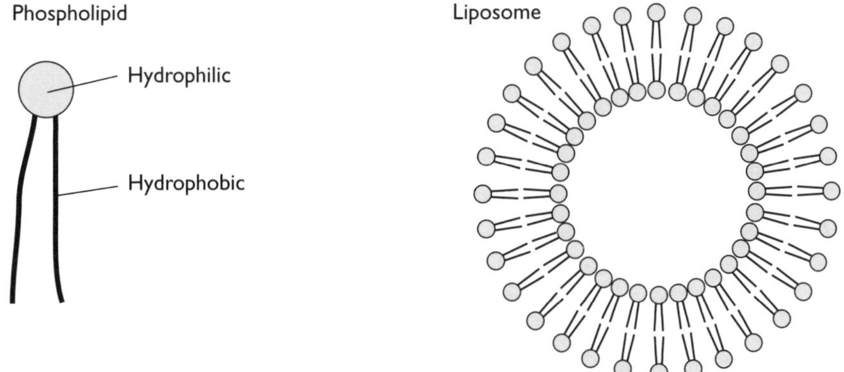

Figure 20.36 Structure of a liposome

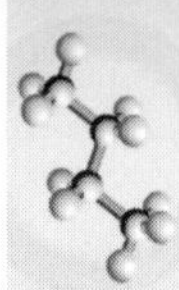

Another method for protecting a drug when it is in the bloodstream is to attach it to a polymer the body will tolerate. One of the most commonly used is 'polyethylene glycol' or PEG:

$$HO—(CH_2—CH_2—O)_n—H$$

If the length of the chain is not too long, PEG is soluble in water (and hence in blood). It does have the disadvantage that a molecule of PEG can carry only two drug molecules.

Drug delivery

Although the synthesis of new drugs is a major challenge, an equally challenging aspect of medicinal chemistry is to make sure the drug is delivered to its target. Many drugs are swallowed and so have to pass through the body's digestive system in order to get into the bloodstream and be carried to the target site. Acid conditions in the stomach, mildly alkaline ones in the small intestine and the presence of a range of enzymes to digest food can potentially break down or change the structure of a drug.

Things to remember

- Drugs have to be designed to be effective and to have minimal side effects.
- Drugs act by having the right shape to fit onto the receptor and the correct binding sites.
- Many drug molecules are chiral (exist as optical isomers) and only one of the forms is likely to have the desired effect.
- If only one of these forms can be synthesised (asymmetric synthesis) it saves time and resources.
- Natural products are often useful starting materials for new drugs.
- Liposomes and polymers (such as PEG) can be used to protect drugs from being broken down and deliver them to their target.

Properties of polymers

We have already covered a number of aspects of polymers in Chapter 19. This section is more concerned with the properties of polymers and how they might be modified for particular purposes.

Addition polymers

The properties of addition polymers can be modified in number of ways, based on their structure and the bonds between the polymer strands. Addition polymers containing only carbon and hydrogen have only temporary dipoles between the strands, and as such are easily deformed. The longer the chain, the stronger are the van der Waals forces between adjacent chains. The presence of side chains also affects the bulk properties of an addition polymer. Unbranched polymers pack closer together than polymers with lots of side chains and this increases the bulk density of the polymer. These differences lead to two distinct forms of poly(ethene) — LDPE (low density poly(ethene)) that has lots of side chains (Figure 20.37(a)), and HDPE (high density poly(ethene)) in which the polymer strands are packed more closely together (Figure 20.37(b)).

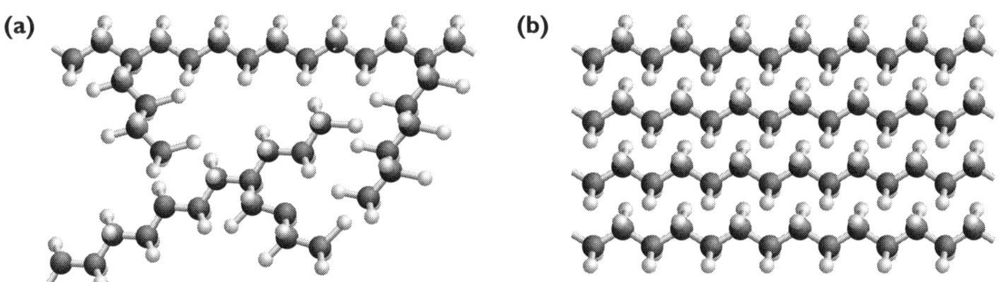

Figure 20.37(a) Structure of (a) LDPE and (b) HDPE

Condensation polymers

In condensation polymerisation the nature of the monomers can vary (in proteins for example) with only the bonds formed between the monomers being constant. This means that the properties of a polymer can be varied by changing one or both of the monomers. Nylon-6,6, proteins and Kevlar® are all polyamides, yet they have very different properties:

- Nylon-6,6 is made from two monomers with a similar hydrocarbon chain. It consists of long chains suitable for use in textiles.
- Proteins are made from a variety of amino acids and have a huge range of properties depending on the amino acid sequence. Some, for example enzymes, are water soluble; others, for example collagen, are tough and water-resistant.
- Kevlar® is made from two monomers with similar hydrocarbons in the chain (but much bulkier than in nylon-6,6). It is extremely tough and is used in the construction of bulletproof vests, and in high performance tyres.

Other polymers have been developed that conduct electricity or emit light, opening up the possibilities of 'wearable' MP3 players, mobile phones and computers.

Things to remember

- Addition polymers formed from hydrocarbons tend to deform because they are held together by van der Waals forces between the chains. These are dependent on chain length.
- Bulk properties of addition polymers depend on the packing of the chains, which in turn depends on how branched they are. Those with more branching have lower density and are generally weaker.
- Addition polymers that contain conjugated double bonds have the ability to conduct electricity, or emit light.
- Condensation polymers have a wide variety of properties, depending on the structure of their monomer(s).
- Most natural polymers, such as proteins, are condensation polymers.

Nanotechnology

A basic definition of **nanotechnology** is the engineering of functional systems on the molecular scale. It is a science that had its beginnings in the late 1970s and early

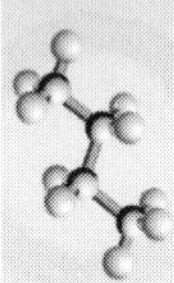

1980s and, in general, it refers to systems on a scale of 1 to 100 nanometres (1×10^{-9} to 1×10^{-7} metres) — the average human hair is about 25 000 nanometres wide.

Some of the early work on nanotechnology centred on a form of carbon known as 'buckminsterfullerene' or a 'buckyball' (Figure 20.38), discovered in 1985.

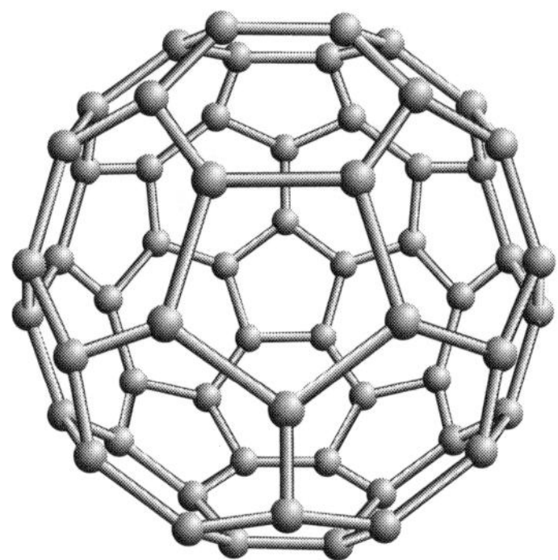

Figure 20.38 A 'buckyball'

The discovery of a new form of a pure element is a rather rare occurrence, particularly for a common element. Therefore, reports of such discoveries generate an unusual amount of excitement among scientists. A whole new chemistry has developed in which fullerene molecules are manipulated to form compounds. Since the C_{60} sphere is hollow, other atoms can be trapped within it.

Linked to this discovery was the fact that single sheets of carbon atoms could be isolated from graphite. This substance, known as 'graphene', consists of one-atom-thick layers of carbon atoms arranged in hexagons (Figure 20.39).

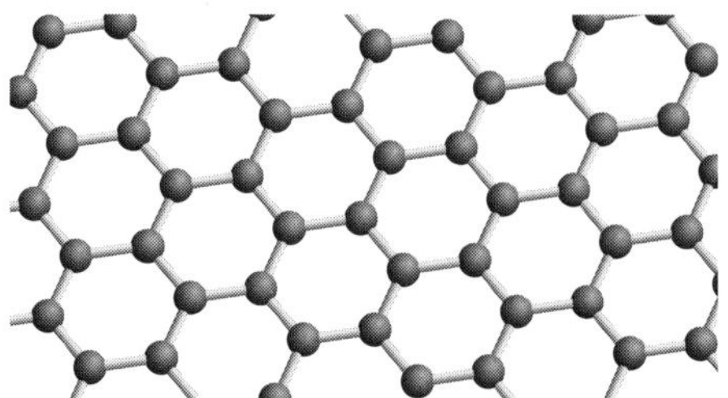

Figure 20.39 Structure of graphene

Transistors made from such samples have been shown to operate at gigahertz frequencies — comparable to the speed of modern computers. The material could theoretically operate near terahertz frequencies, hundreds of times faster.

In addition to this, these sheets can be rolled into cylinders known as **nanotubes** (Figure 20.40) and even combined with half a buckyball to produce nanoscale test tubes.

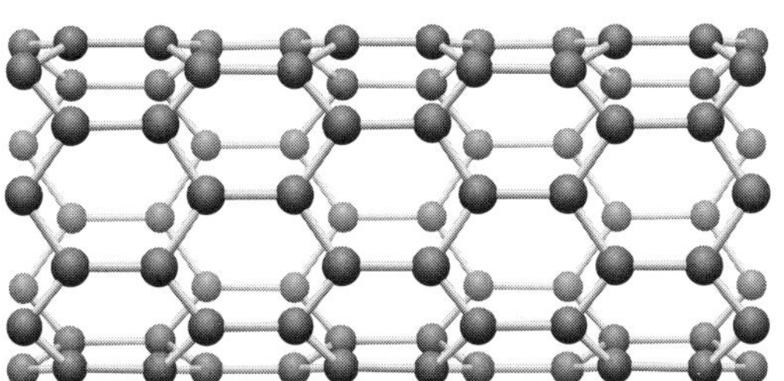

Figure 20.40 A nanotube

The range of applications of nanotechnology changes almost every month, but some established nanotechnologies that you might be asked about are:

- Modern sunscreens contain a variety of compounds to absorb ultraviolet light. Some sunscreens contain nano-sized particles of titanium dioxide, which reflect ultraviolet light. Traditional sun-block preparations contain titanium dioxide, but the size of the particles mean that visible light is reflected and the sunscreen is white. The nano-size particles appear colourless.
- Manufacturers of inkjet paper now include a layer of ceramic nanoparticles in the paper. These protect the image 'fixed' in lower layers with the result that such images may have a lifetime of up to 100 years.
- Coatings of nanoparticles on windows let light in while keeping heat out. This is achieved by the gaps between the nanoparticles being smaller than the wavelength of infrared radiation, but not so small as to exclude visible light.
- There has been considerable development over the past 5–10 years in the use of nanoparticles for cosmetic products including moisturisers, haircare products, make-up and sunscreen (discussed earlier). Liposomes and niosomes (which have different units in the spherical structure) are used in the cosmetic industry as delivery vehicles, particularly for chemicals used to hydrate and prevent aging of the skin. Such chemicals include anti-oxidants such as vitamins A and E.

Things to remember

- Nanotechnology is based around particles 1–100 nm in size.
- The properties of nanoparticles can be very different from the bulk properties of the substance concerned — for example, graphene is transparent whereas graphite is grey-black.
- Nanotechnology has a large and growing range of applications in a wide variety of contexts.

Environment and energy

This section of the syllabus looks at the pollution of the environment and the part that chemistry can play in removing pollution, the efficient use of raw materials and the development of new, efficient energy sources.

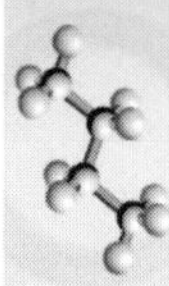

Pollution

Oil

One of the major environmental problems in recent years has been pollution due to oil spillages. These have been caused by oil tankers sinking or running aground or by leakages from oil wells and pipelines. They are usually problems of the marine and coastal environments. The oil floats on the surface of the water and is washed ashore where it causes immediate and longer-term damage to aquatic bird and fish populations. If the oil can be prevented from reaching shore, the damage is likely to be much reduced.

One method of trying to prevent oil slicks from spreading is to use booms made of porous materials called 'sorbents' that are able to soak up the oil. Traditional booms also absorb water and soon sink as they become waterlogged. More recently, new materials have been developed to act as sorbents. These are able to absorb more than 200 times their own mass of oil without sinking. The key has been the creation of a fibreglass sorbent that has fluorinated hydrocarbons trapped in the structure. The hydrocarbon is hydrophobic ('water-hating') and so repels the water while allowing the oil to be absorbed.

Polluted soil

Land that has been used for industrial activity may become polluted with materials related to the industry. There is a huge variety of these pollutants, including heavy metals. The removal of harmful contaminants from soil is known as **soil remediation**. There are a number of methods that can be used, the choice of which depends on the nature of the contaminants.

Excavation and dredging are the most common processes used in soil remediation. These processes involve extracting soil that is contaminated and judged to be unrecoverable using current technology, and transporting it to a landfill site set aside for this purpose. Purified soil may be used to fill in the area where the extraction took place.

Soil remediation is sometimes accomplished using a process known as pump and treat. This approach involves the removal of contaminated groundwater, followed by various methods to purify the extracted liquid. The soil is also extracted and treated to remove various contaminants. It is then returned to its original position. The purified water is pumped back into the purified soil, effectively restoring the ecological balance of the area.

Where a high concentration of contaminants is present in groundwater, a process known as **solidification** is used. Chemical reagents that will combine with the materials in contaminated water and sludge are added to make solid compounds that can be separated from, or filtered out of, the water. This way the water is left in the natural ecosystem but the contaminant chemicals are removed.

Bioremediation involves the addition of specific bacteria or plants that feed on the contaminated particles and create harmless by-products. These organisms must be chosen carefully as they may multiply rapidly with no natural predators.

Bioremediation is more common in contaminated water, but can work in soils. Naturally-occurring microorganisms in soil can break down some organic contaminants into carbon dioxide and water, providing that other nutrients are supplied. These can be pumped deep into the soil in calculated amounts. This technique has been used successfully to treat contamination by a wide range of organic compounds.

The use of plants in removing contaminants is sometimes referred to as **phytoremediation** and is not a new process. Certain plants (**hyperaccumulators**) are able to accumulate, degrade or make harmless some contaminants in soils, water or air. Metals, pesticides, solvents, explosives, crude oil and its derivatives have been treated in this way. Mustard plants, alpine pennycress and pigweed have proven to be successful at removing contaminants. Water hyacinth is highly successful at removing both heavy metals and cyanide from water supplies.

Pollution in the atmosphere

We saw in Chapter 15 that chlorofluorocarbons pose a significant threat to the Earth's ozone layer and contribute to global warming. As a result, they have been largely phased out of commercial (and some industrial) use by the finding of alternatives. CFCs are not the only compounds that threaten the ozone layer or contribute to global warming, but you will not be required to remember specific details of others.

In the past, vehicle exhausts have caused considerable pollution of the air at street level. The phasing out of lead additives and the increasing use of catalytic converters has largely solved this problem in developed countries. However, large amounts of carbon dioxide are still being produced that add to global warming.

Recent developments in the use of electricity and hydrogen to power vehicles may help to solve this problem (see 'New forms of energy') but at present the technology is expensive.

Extending the life of resources

Increasing awareness of both pollution and the limited resources on Earth has led to innovation in chemical research and the introduction of the phrase **green chemistry**.

Among the changes in outlook of research, the following principles are now widely adopted:

- prevention of waste (cheaper than cleaning up afterwards)
- energy efficiency is maximised
- feedstocks for processes should be sustainable
- chemical products should break down naturally
- hazardous solvents should be avoided

Copper mining

Examples of this in practice include copper mining. At the beginning of the twentieth century the world's annual demand for copper was around half a million tonnes, and ores containing about 4% copper were the limit of what was regarded as profitable. Today, the demand for copper is approaching 5 million tonnes per year and ores

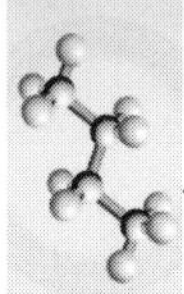

once regarded as worthless, and containing less than 1% copper, are now being used for extraction. Not only are companies mining new ores, but they are also re-working 'spoil heaps' from previous extractions.

A new, energy-efficient technology called **bioleaching** has been developed to enable this. Most copper minerals are sulfides, with chalcopyrite ($CuFeS_2$) being the most abundant and thus the most important economically. Sulfide minerals are insoluble in water and acid solutions unless they are first oxidised. While exposure to air is sufficient to oxidise these minerals, the process is slow and inefficient. The kinetics of the oxidation process are vastly improved by the introduction of *Thiobacillus ferrooxidans* and *Thiobacillus thiooxidans* bacteria to the system.

The overall chemical reactions for chalcopyrite are:

$$4CuFeS_2 + 11O_2 + 6H_2O \xrightarrow{\textit{Thiobacillus ferrooxidans}} 4CuSO_4 + 4Fe(OH)_3 + 4S$$

$$2S + 3O_2 + 2H_2O \xrightarrow{\textit{Thiobacillus thiooxidans}} 2H_2SO_4$$

Similar technology is being used to extract other metals — for example gold, nickel and zinc.

Supercritical carbon dioxide as a solvent

A supercritical fluid is a gas that is compressed and heated to have both the properties of a liquid and a gas at the same time. Carbon dioxide becomes supercritical at a pressure of about 7300 kPa and a temperature of 31 °C.

Supercritical carbon dioxide is well established as a solvent for use in extraction processes. There are number of reasons for this. It can penetrate a solid sample faster than liquid solvents because of its high diffusion rate, and can rapidly transport dissolved solutes from the sample matrix because of its low viscosity. There are also fewer solvent residues present in the products.

One of the commonest uses of supercritical carbon dioxide is decaffeination of coffee and tea, which is now an established industrial process. Other examples have been developed, including the extraction of hops, natural products, high-value pharmaceutical chemicals, essential oils and environmental pollutants. The recovery of the solvent is straightforward and it can be re-used.

New forms of energy

Traditional forms of energy have relied on fossil fuels either directly or indirectly (as in electricity generation). These are generally inefficient (rarely better than 35–40%) and are also a source of carbon dioxide (and possibly other pollutants). In addition, a number of the materials used for fuels also have important uses as feedstocks — for example crude oil for polymers and raw materials for the organic chemistry industry. Converting from coal-burning to gas-powered power stations increases the efficiency of energy conversion to electricity, and if a 'combined heat-and-power' system is used conversion rates approaching 80% can be achieved.

There are, however, alternatives to burning valuable resources such as oil.

Biofuels

Bioethanol produced by fermenting agricultural products can be used to supplement petrol and/or gasoline. This has been used for a number of years in parts of Africa and South America. More recently developments have begun in the USA and Europe. Another biofuel is biodiesel, which has been developed from a number of oilseed crops. Biofuels are potentially a useful replacement for fuels based on crude oil, but their production needs careful analysis in terms of the overall energy balance — for example, does the production of the crop need more fertiliser produced using fossil fuels?

Batteries and fuel cells

Battery technology has undergone a revolution in recent years with the development of alternatives to traditional alkaline cells and of small rechargeable batteries that can be recharged more than 1000 times. This saves waste and pollution since batteries are no longer thrown away once their power has been used. Many stores now collect used batteries. Rechargeable lead–acid batteries have been used in cars for many years and are still in use. Recharging the batteries still requires electricity generated by conventional means, although solar rechargers are now available.

Fuel cells are an alternative to batteries, although for large-scale uses the technology is still in its infancy. In a fuel cell, a fuel (such as hydrogen or a hydrocarbon gas) and oxygen are passed over electrodes. This generates electrical power. Such systems were developed originally for spacecraft, but more recently car manufacturers have shown interest in this technology with a number of prototypes being developed.

Hydrogen

Hydrogen is a useful fuel in that the product of combustion, water, is non-polluting. However, hydrogen is not a naturally-occurring fuel. It has to be produced from either existing feedstocks, such as natural gas, oil and coal, or from renewable sources such as water by electrolysis. Even if this could be achieved using renewable energy such as solar energy there are still practical problems to be overcome.

One of these problems is storage of the hydrogen. It has to be compressed so that it can be stored in a small enough space to fit a vehicle, and this presents the risk of an explosion in an accident. Two solutions to this problem are being researched:

- absorbing the hydrogen onto a solid material that can then be released chemically as needed
- using nanotechnology to produce a material with molecular-sized holes to absorb the hydrogen and release it as required

In the last few years, work on buckyballs has led to hopes that they might be useful as a hydrogen-storage medium. Two strategies are under development — one involving

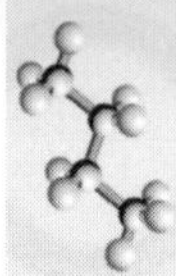

coating the surface with Group I or Group II metal atoms and the other by creating buckyball cages to contain hydrogen (Figure 20.41).

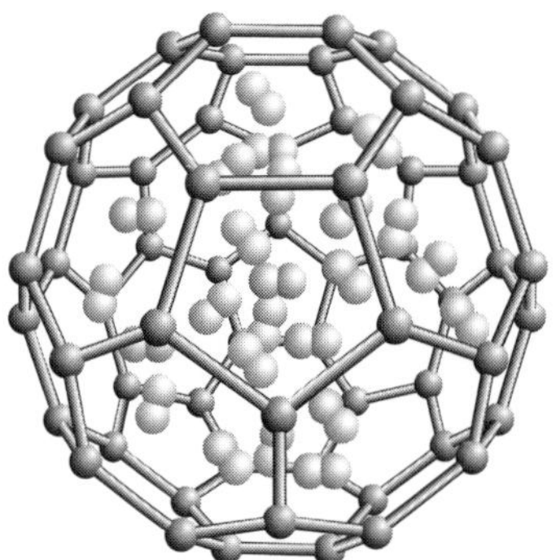

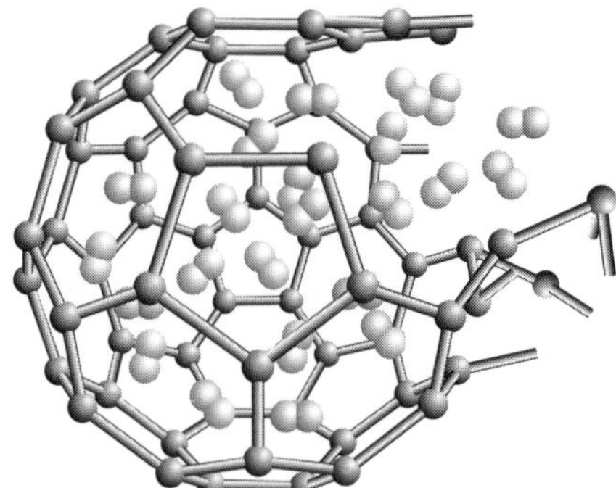

Figure 20.41 Buckyball cages

Things to remember

- The development of efficient and affordable solutions to environmental pollution depends on knowledge of the physical and chemical properties of the pollutants.
- It is better to reduce waste by efficient chemical processes than to clean up pollution afterwards.
- There are limited resources on Earth and efficient use of these, including recycling, is critical to our consumer economy.
- Chemistry has a vital part to play in the development of new energy supply and the more efficient use of existing fossil fuels.

Experimental skills and investigations

This section is needed for both the AS and the A2 examinations. The AS practical component Paper 3 and the A2 practical component Paper 5 are covered separately.

Paper 3: AS practical paper

Skills tested on this paper are:

- manipulation of apparatus
- presentation of data
- analysis and evaluation

Almost one-quarter of the total marks for the AS examination are for experimental skills and investigations. These are assessed on Paper 3, which is a practical examination worth 40 marks. Although the questions are different each year, the number of marks assigned to each skill is always approximately the same. This is shown in the table below.

Skill	Minimum mark allocation*	Breakdown of marks	Minimum mark allocation*
Manipulation, measurement and observation, MMO	12 marks	Successfully collecting data and observations	8 marks
		Quality of measurements or observations	2 mark
		Making decisions relating to measurements or observations	2 marks
Presentation of data and observations, PDO	6 marks	Recording data and observations	2 marks
		Displaying calculations and reasoning	2 marks
		Data layout	2 marks
Analysis, conclusions and evaluation, ACE	10 marks	Interpreting data or observations and identifying sources of error	4 marks
		Drawing conclusions	5 marks
		Suggesting improvements	1 mark

*The remaining 12 marks will be allocated across the skills in this grid, and their allocation may vary from session to session.

The examination consists of two or three questions. One question is an observational problem. You will be asked to carry out particular experiments to investigate one or more unknown substances. These substances may be elements, compounds or mixtures. You could be asked to construct tables to record your observations, analyse your results and draw appropriate conclusions.

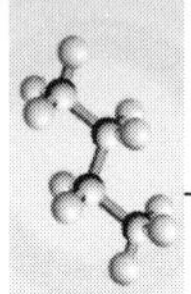

The other question or questions are quantitative — in other words, involve measurement. These could include titrations (volumetric analysis) or measurement of a quantity such as enthalpy of reaction or reaction rate. You will be expected to:

- construct tables, draw graphs or use other appropriate ways of presenting data
- analyse the data
- perform calculations
- draw conclusions

One or more of the questions will ask you to identify sources of error and make suggestions for reducing these.

A detailed breakdown of examiners' expectations for each mark category is given in the syllabus and it is important that you read through this.

Manipulation, measurement and observation

In the examination, unless you are told differently by your teacher or supervisor, you can tackle the questions in any order. It makes sense to attempt the question carrying the most marks first, in case you run out of time. However, with some thought and careful planning this should not happen.

Manipulation and measurement

It is important to think about the accuracy of the different readings you are asked to make. For example, an electronic stopwatch might measure to the nearest one-hundredth of a second, but accuracy to the nearest second is more appropriate in any time measurement you are likely to make. You need to think about the number of decimal places or significant figures to use when you are recording data. It is important to remember that different pieces of apparatus will give measurements with different degrees of accuracy. Overall, any experimental data are only as accurate as the *least* accurate measurement. It is important to understand the difference between accuracy and precision. The diagrams below may help.

	Accurate	**Inaccurate (systematic error)**
Precise		
Imprecise (scattered results)		

You may be asked to use a burette to add a liquid reagent and measure the quantity used. You should have had lots of practice during your AS course at using a burette and reading the scale.

Tip You should read a burette to the nearest 0.05 cm^3, in other words halfway between two of the smallest divisions. You cannot estimate more precisely than this because the smallest drop a burette can deliver is approximately 0.05 cm^3.

A common mistake students make is in calculating the average burette reading used in a titration. This cannot have a greater degree of accuracy than the burette can produce. Therefore, the average burette reading has to end in .00 cm^3 or .05 cm^3. Anything else is wrong and will be penalised.

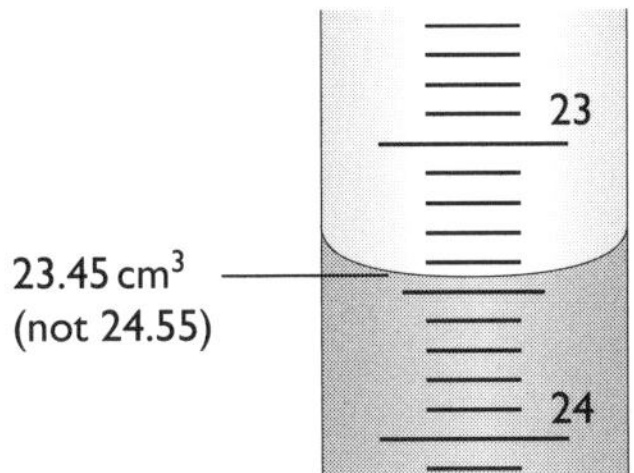

In a titration you need to carry out sufficient repeat titres to establish the 'correct' end point. For a titration with a sharp end point this should be within 0.10 cm^3.

Tip Once you have established the approximate end point, say 22.60 cm^3, for the accurate titrations, you can run in 20 cm^3, swirling the mixture, before adding a few drops at a time. This will save time.

It is important that you always try to pick the same point for determining the end point. If it is a potassium manganate(VII) titration this is when the mixture just has a permanent pale pink colour. If it is an acid–base titration, the end point is when the indicator just remains in its acid state on swirling.

Tip Do not add more than two or three drops of indicator unless instructed otherwise. Using more indicator does not give a more precise end point because all of the indicator has to change to its acid form.

If you are asked to perform a kinetics experiment, read carefully through the instructions before you start. Make sure that you have everything ready before you start the reaction — reagents measured out, stopwatch ready, thermometer in place (if relevant). Being organised is the best way to get good results and to avoid wasting time.

Tip In a kinetics experiment, if you are adding a quantity of one reagent to another, make sure that you start the stopwatch at the same point each time. This could be halfway through adding the reagent, or at the point when the last of the reagent has been added.

If you are carrying out an enthalpy of reaction experiment you will probably be asked to measure the temperature of the reaction mixture every 30 seconds or every minute. It is important that you are organised and ready to do so.

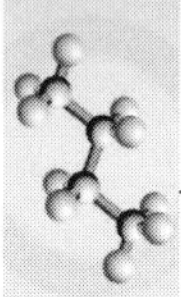

Tip Stir the mixture continuously with the thermometer to make sure that you are measuring the temperature of the whole mixture (this is particularly important if the mixture contains a solid). Get ready by checking the position of the mercury in the thermometer as you are coming up to the next timed reading.

Observations

In most examinations one specific question involves making careful observations, recording them and then drawing conclusions from these observations. If you are to score good marks on this question it is *vital* that you make good observations and record these accurately.

To make sure your observations are correct you must be certain that you follow the instructions completely.

Tip Examples where students make errors include the following:

- adding excess reagent when told to add a few drops *followed* by an excess — this often means that an observation which carries marks is missed
- adding reagents in the wrong sequence — this can mean that changes which should be observed do not happen
- using too much of the 'unknown' or too much reagent — this can mask observations
- using a solid 'unknown' rather than a solution of the substance —observations may be different or masked
- failing to test for a gas produced (lighted splint, glowing splint, indicator paper, smell) — marks will be lost because of observations missed

Presentation of data and observations

It is important that, having carried out a practical task, the results are recorded systematically and logically. In some cases, such as observational exercises, there may be a table printed on the exam paper for you to fill in. In other exercises, such as titrations, you may need to draw your own table for the data.

You may also be asked to draw a graph to display the results of an experiment. Here are some tips on what is needed to score high marks when drawing a graph.

Tip There are a number of key features on a graph that examiners look for:

- a title describing what the graph represents
- both axes labelled, including the correct units
- the independent variable plotted on the x-axis and the dependent variable on the y-axis
- use of a sharp pencil for plotting the points and sketching the graph
- a sensible scale chosen to make the most of the graph paper (at least half in both dimensions)
- drawing a line of best fit for data that vary continuously
- identifying anomalous results (outliers) and not giving these undue weight

The diagram below shows how to draw a 'line of best fit'. The line is placed so that points are equidistant from the line with equal numbers of points on each side.

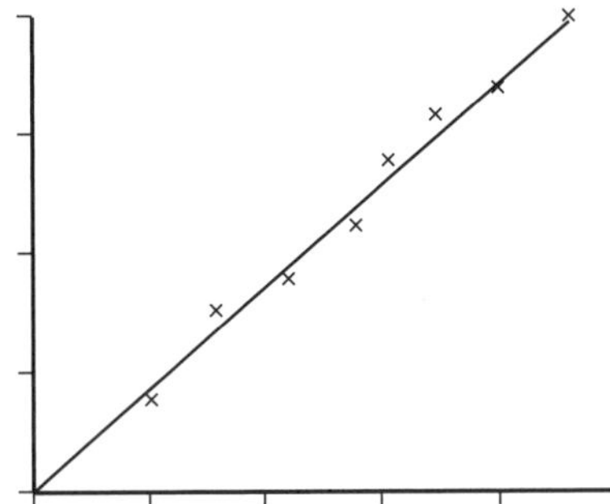

The diagram below shows what is meant by an 'outlier' — a result which, for some reason, is clearly not part of the data set we want to use.

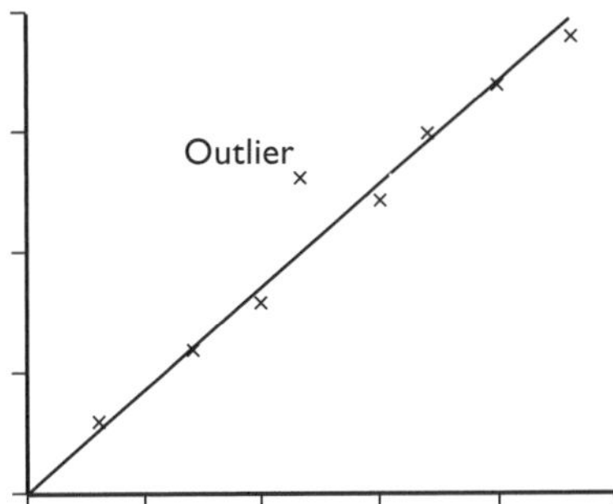

When an experiment produces qualitative data, it is just as important to represent the data accurately. Accurate recording of what happened at different stages of an experiment — for example when a few drops of reagent were added, when more reagent was added and when the reagent was present in an excess — will help you explain what is happening and score the available marks.

Try to use accurate and unambiguous language when describing colours or colour changes. Make sure that it is clear whether you are referring to a solution or a precipitate in the solution.

Tip The words you use must be clear to the examiner as well as to yourself.

- Do *not* use words such as 'see-through' or 'transparent' when you mean 'colourless'. It is perfectly possible to 'see through' potassium chromate(VI) solution but it is not colourless, and copper(II) sulfate solution is 'transparent', but it is not colourless.
- Try to be as precise as possible about colours. A simple word like 'blue' is rarely enough to describe colour changes such as those that take place when aqueous ammonia is progressively added to a solution of copper(II) ions.
- If a gas is given off remember that the observation is that 'bubbles are produced' or 'effervescence takes place', *not* just that 'carbon dioxide is evolved'. If the gas

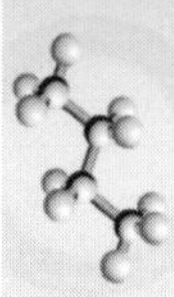

is carbon dioxide you also need to say how you tested for it — for example 'it turned limewater milky/cloudy'.

Analysis, conclusions and evaluation

Analysis and conclusions

In order to draw conclusions and evaluate the procedures you have carried out, you first need to analyse the results of any practical you complete, and then explain what they show. Before reaching a conclusion you may need to carry out a calculation. If this is the case, you should show the key steps in the calculation so that it can be followed by the examiner who can then check the accuracy. For example, in a titration a series of repeats is undertaken to ensure accuracy and to identify any anomalous results. When the average titre is calculated, an anomalous result should be excluded from the calculation.

You may be asked to draw a graph to display results because this is an excellent way to show trends or relationships. A straight-line graph passing through the origin shows a directly proportional relationship between the dependent and independent variables. Sometimes the graph will be a curve and you may need to measure the gradient of the curve at different points. When doing this, use large triangles to calculate the gradient as this gives a more accurate calculation. Remember that a graph provides evidence to support a conclusion.

Evaluation

This is one of the hardest skills to develop. It relies on your ability to think critically about the reliability of the data you have collected and the conclusions drawn.

One way to do this is to think about the errors that affect results. In any experiment there are two types of error:

- Random errors cause results to fluctuate around a mean value. The results are made more reliable by taking a number of readings and averaging them.
- Systematic errors affect all measurements in the same way, giving values higher or lower than the true result. Hence, systematic errors cannot be averaged out. An example of this is the heat loss in an enthalpy change experiment.

Remind yourself of these differences by looking back at the diagram on p. 220.

Tip When you evaluate an experiment you should be able to do the following:

- suggest improvements to the procedures you used
- compare repeated results to judge their reliability
- identify any anomalous results (outliers)
- identify variables you need to control
- estimate uncertainty in measurements
- distinguish between random and systematic errors

Paper 5: A2 assessment

Skills tested on Paper 5 are:
- planning
- analysis and evaluation

Practical skills and investigations are examined on Paper 5, which is worth 30 marks. This is *not* a laboratory-based examination, but it tests the practical skills that you will have developed during your A-level course. The table below shows the breakdown of marks.

Skill	Total marks	Approximate breakdown of marks	
Planning	15 marks	Defining the problem	5 marks
		Methods	10 marks
Analysis, conclusions and evaluation	15 marks	Dealing with data	8 marks
		Evaluation	4 marks
		Conclusion	3 marks

It is expected that you will know how to:
- plan how to carry out an experiment
- perform the experiment according to your plan
- evaluate what you have done

The examination consists of two or more questions. One of these will ask you to design an experimental investigation of a given problem. This question is *not* structured. It requires you to answer using diagrams, extended writing, flow charts, tables and equations. You may also be asked to express a prediction in the form of a written hypothesis that links the independent and dependent variables or in the form of a graph showing the expected outcome.

There may be questions that contain some experimental data and which ask you to analyse, evaluate and draw conclusions from those data. Once again this type of question is *not* highly structured and you will need to decide for yourself the methods you use to tackle the question.

One or more questions may be on areas of chemistry that are difficult to investigate experimentally in a school laboratory. Such questions will *not* require knowledge of equipment or theory beyond that in the A-level syllabus. Any information you need that you would not be expected to know will be given in the question.

A detailed breakdown of examiners' expectations for each mark category is given in the syllabus, and it is important that you read through this.

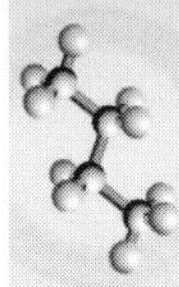

Planning

Planning can be a difficult practical skill to master. It depends on thinking carefully about the practical problem set. It also relies on a good understanding of the practical experience you have gained during the course. The following may make it easier for you to gain marks.

Tip The writing of a plan is best tackled in stages. Practical exercises vary greatly in their demands, so it is probably best to consider two types of plan — one for quantitative exercises and another for qualitative exercises.

(1) Quantitative exercises

- You should be able to identify the independent and dependent variables from the information given.
- You should be able to identify other key variables and propose measures to control these.
- You should be able to make a quantitative prediction of the likely outcome of the experiment (if required).
- The plan needs to be able to test any prediction/hypothesis in a reliable, unambiguous and reproducible way.
- Even if the question does not ask for a formal prediction/hypothesis, you should have a clear idea of what you expect the results to show.
- The data provided in the question will need processing in some way to enable analysis and evaluation (see later). This also means that some discussion of how this processing is to be carried out needs to go into the plan.
- Any recording, graphical and numerical processes should be stated.
- The steps by which the experimental procedure and analysed data will be evaluated should be included.

(2) Qualitative exercises

This type of exercise might involve planning the preparation of a given mass of a compound or an analysis scheme for an unknown compound. The plan should be sufficiently detailed that, if the experiment were performed by a competent chemist, it would produce the anticipated outcomes. Many students lose marks at this stage by not producing a sufficiently detailed plan and relying on the examiner to 'fill in the gaps'.

- The suggested plan must be workable given the apparatus available, and the scale required (e.g. for producing a known mass of compound).
- The quantities of reagents to be used should be specified.
- The heating or cooling of the reaction mixture and the method chosen to do so should be included.
- The sequence of carrying out tests on an unknown compound should be stated so that false results can be excluded.
- Any purification techniques employed in the production of a compound should be included.

Analysis, conclusions and evaluation

Analysing data

In analysing data, you will need to be able to use your understanding of the theory behind a given experiment. Some people find it difficult to put together a clear, reasoned and justified argument to support this. The stages in the argument must be clear and easy for the examiner to follow.

Tip When analysing numerical data you need to be proficient in handling the mathematics involved and confident in carrying out the calculations needed — including the correct use of significant figures. Remember, the number of significant figures to which an answer is expressed shows the precision of the measured quantities. The general rule is that you should use the same number of significant figures as are found in the *least precise* measurement.

You also need to be able to analyse errors in the experiment. These fall into two groups — those associated with the use of a particular piece of apparatus and those linked to the level of competence of the operator or flaws in the procedure. One way to start thinking about apparatus errors is to compare the use of a burette in measuring different volumes of a liquid. If you use it to measure out $5\,cm^3$, the error in measurement will be five times greater than if you use it to measure $25\,cm^3$.

Evaluation

This is one of the harder skills to develop because to be successful in evaluating an experiment or procedure you need to have a clear idea of the aims, objectives and predicted outcome.

Evaluation can include:

- identifying anomalous results (or outliers)
- deducing possible causes for the anomalous results and suggesting ways of avoiding them
- a view of the adequacy of the range of data used
- commenting on the effectiveness of measures taken to control variables
- an argued judgement on the confidence you have in the conclusions reached

It goes without saying that if anomalous results are to be identified then the expected pattern of results must be known. This could be from a prediction as part of a hypothesis about the experiment, or because an experimentally determined point does not fit the trend of other data. Having identified the anomaly you should suggest what might have caused it, and have a strategy for dealing with it (which might involve excluding it or repeating the measurement). The effect of including and excluding an outlier in drawing a line of best fit is shown in the diagram on the next page.

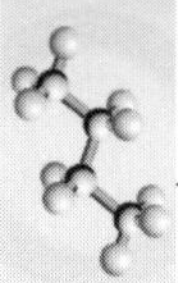

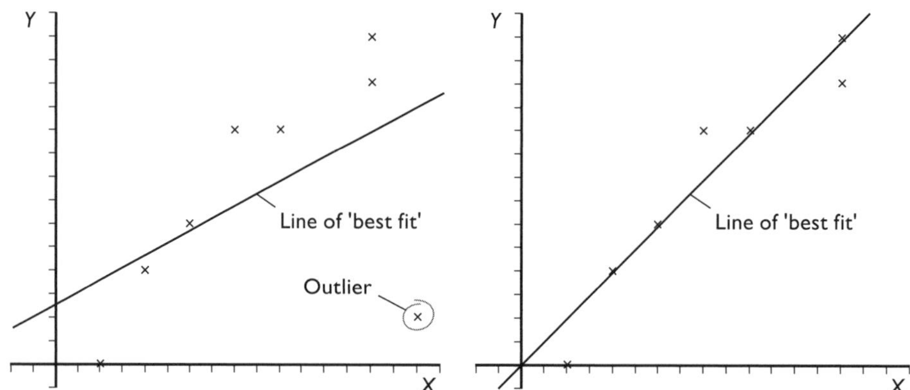

Tip So what should you do if the results of a quantitative experiment are inaccurate? The first thing is to establish whether the errors are due to the apparatus used or to the data collected.

If the errors in the data collected exceed the apparatus errors then it is important to identify the flaws in the procedure which led to these errors and to suggest a more reliable procedure. If the error is a result of a temperature fluctuation in the laboratory, you might suggest using a water bath with a thermostat to reduce this fluctuation. If the errors in the data do not exceed those due to the apparatus used, you need to suggest how to reduce the apparatus errors. If one of the sources of error is in weighing a solid, it is not sufficient to say 'use a better balance'; you must quantify this — for example, 'use a balance accurate to ±0.01 g'.

Remember that there is no credit for saying that the experiment went well or gave good results. You *must* say why, and give evidence to support this.

Drawing conclusions

This is usually the final stage in commenting on an experiment or procedure. It relies on you having a thorough knowledge of the chemistry involved.

The conclusion you draw must be based on how well the data collected matches the original hypothesis or prediction, or is supported by the results of qualitative tests. You will be expected to use your knowledge of the theoretical background to the experiment or procedure and to make judgements about the data or results based on this knowledge.

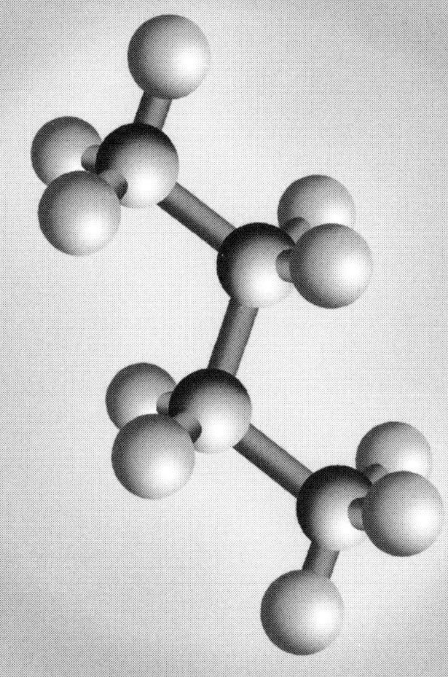

Questions & Answers

In this section are two sample examination papers — one similar to the Cambridge International Examinations (CIE) AS Chemistry Paper 2, and the other similar to the CIE A2 Chemistry Paper 4. All of the questions are based on the topic areas described in the previous parts of the book.

You have 1 hour and 15 minutes to do the AS paper. There are 60 marks, so you can spend just over one minute per mark. You have 2 hours for the A2 paper. There are 100 marks, so again you can spend just over one minute per mark. The A2 paper also contains questions on the applications part of the syllabus, and these may take you a little longer. If you find you are spending too long on one question, then move on to another that you can answer more quickly. If you have time at the end, then come back to the difficult one.

It is important to remember that when you take an A2 paper it is assumed that you already know the AS material. Although the questions in the paper are focused on the A2 content from the syllabus, the underlying chemistry may be based on work covered at AS.

Some of the questions require you to recall information that you have learned. Be guided by the number of marks awarded to suggest how much detail you should give in your answer. The more marks there are, the more information you need to give.

Some of the questions require you to use your knowledge and understanding in new situations. Don't be surprised to find something completely new in a question — something you have not seen before. Just think carefully about it, and find something that you do know that will help you to answer it. Make sure that you look carefully at the information provided in the question — it will have been included for a reason!

Do think carefully before you begin to write. The best answers are short and relevant — if you target your answer well, you can get a lot of marks for a small amount of writing. Don't say the same thing several times over, or wander off into answers that have nothing to do with the question. As a general rule, there will be twice as many answer lines as marks. So you should try to answer a 3-mark question in no more than six lines of writing. If you are writing much more than that, you almost certainly haven't focused your answer tightly enough.

Look carefully at exactly what each question wants you to do. For example, if it asks you to 'Explain', then you need to say *how* or *why* something happens, not just *describe* what happens. Many students lose large numbers of marks by not reading questions carefully.

Following each question in this part, there is an answer from Candidate A, who might achieve a C or D grade, and an answer from Candidate B who might achieve an A or B grade. The candidates' answers are followed by an examiner's comments. You might like to try answering the questions yourself first before looking at these.

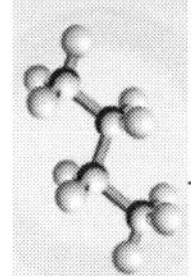

AS exemplar paper

Question 1

The first six ionisation energies of an element X are given in the table.

Ionisation energy/kJ mol^{-1}					
First	Second	Third	Fourth	Fifth	Sixth
550	1064	4210	5500	6908	8761

(a) Define the term *first ionisation energy*. (3 marks)

(b) Write an equation, with state symbols, for the *third* ionisation energy of element X. (2 marks)

(c) Use the data provided to deduce in which group of the periodic table element X is placed. Explain your answer. (3 marks)

The first ionisation energies of the Group IV elements are given below.

Element	C	Si	Ge	Sn	Pb
1st I.E./kJ mol^{-1}	1086	789	762	709	716

(d) Use your knowledge of the atomic structure of these elements to explain the *trend* in ionisation energies. (3 marks)

Total: 11 marks

Candidate A

(a) It is the energy required to convert a mole of atoms ✗ of an element into a mole of cations, ✗ with each atom losing one electron ✓.

There are two errors in this definition. The candidate fails to refer to the gaseous state of both the atoms and cations.

(b) $X(g) - 3e^- \rightarrow X^{3+}(g)$ ✗ ✓

The candidate has confused the 3rd ionisation energy with the loss of three electrons. The state symbols are correct, for 1 mark.

(c) X is in Group II of the periodic table ✓.

This correctly states that X is in Group II. However, the candidate has not explained the evidence and therefore loses 2 of the 3 available marks.

(d) The atoms are getting bigger so the electrons are further from the nucleus making them easier to remove ✓. The outer electrons are screened from the nuclear charge ✓.

This is a fairly good answer. However, there is no mention of the increasing nuclear charge.

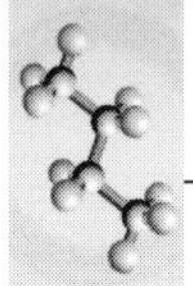

Candidate B

(a) This is the energy needed to remove one electron from each ✓ of 1 mol of gaseous atoms ✓ of an element to form 1 mol of gaseous cations ✓.

This good answer gives all three points in the mark scheme.

(b) $X^{2+}(g) - e^- \rightarrow X^{3+}(g)$ ✓ ✓

The equation is correct, as are the state symbols.

(c) **X** is in Group II of the periodic table ✓. There is a large jump in energy to remove the 3rd electron ✓, which is from a full shell ✓.

This very good answer uses the data in the table and the candidate's own knowledge of the arrangement of electrons in atoms.

(d) There are two effects here. First, the atoms are getting bigger so the electrons are further from the nucleus making them easier to remove ✓. Second, the outer electrons are screened from the nuclear charge reducing its pull on them ✓.

This is a good answer. However, the candidate has not mentioned that these two effects outweigh the increasing nuclear charge and so fails to score the third mark.

Question 2

The modern periodic table is based on one proposed by Mendeleev following his observations of patterns in the chemical properties of the elements.

The diagram shows the first ionisation energies of the first 20 elements in the periodic table.

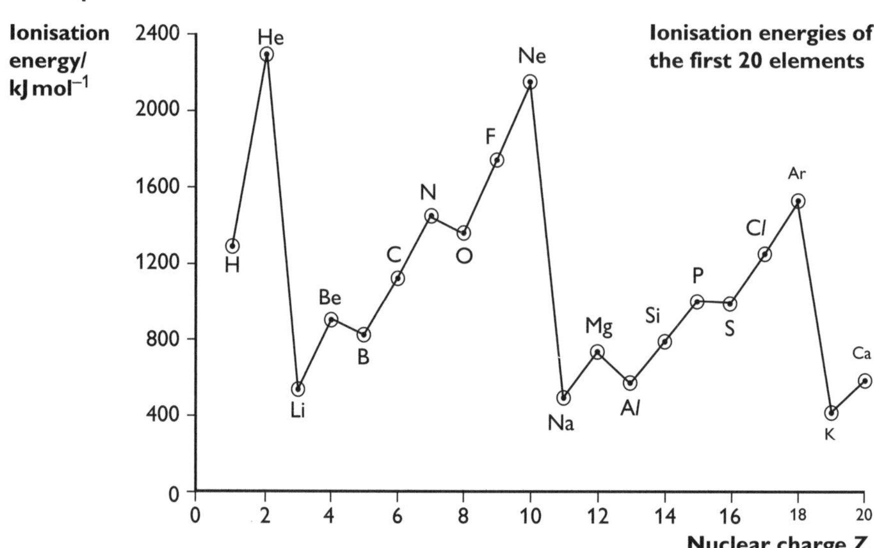

(a) Look at the section of the diagram from Li to Ne. Explain why there is a general increase in first ionisation energy. (3 marks)

(b) Explain why the first ionisation energy of the third element in each period (B and A*l*) is lower than that of the second elements (Be and Mg). (2 marks)

(c) Explain why the first ionisation energy of oxygen is lower than that of nitrogen. (2 marks)

(d) (i) Explain why the first ionisation energy of potassium is lower than that of sodium. (1 mark)

(ii) Describe a chemical reaction that illustrates the effect this has on the reactivity of the two elements. (2 marks)

(e) When Mendeleev produced his table, he did not include the noble gases (Group 0 elements). Suggest why. (1 mark)

Total: 11 marks

Candidate A

(a) For each element an extra proton is added in the nucleus ✓.

It is important to look at the number of marks available for each question. Here there are 3 marks, suggesting you need to make three points. However, this candidate has made only one point and so can gain only 1 mark.

(b) The third elements have electrons in p-orbitals ✓ but the second elements only use s-orbitals ✗.

This answer is partially correct. However, the important point is that electrons in p-orbitals are held less strongly because they are, on average, further from the nucleus.

(c) Nitrogen has one electron in each p-orbital but oxygen has one p-orbital with a pair of electrons ✓.

Again, this is partially correct. The other point is that the paired electrons in oxygen repel one another, making it easier to remove one of them.

(d) (i) The outer electron of potassium is not held as tightly as that in sodium ✗.

This answer is really re-stating the question. The point is that potassium has an extra shell of electrons between the nucleus and the outermost electron.

(ii) With water, sodium fizzes ✓ but potassium burns ✗.

This gives a correct observation for sodium and suggests that the potassium reaction is more vigorous. However, it isn't the potassium that burns, but the hydrogen that is produced.

(e) He didn't know how to make them react ✗.

The answer is that Mendeleev did not know that these elements existed.

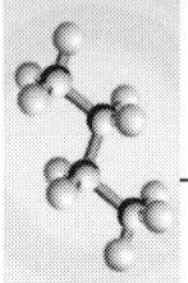

Candidate B

(a) As we go from Li to Ne we are adding an extra proton and an extra electron ✓. All the protons go into the nucleus, increasing the attraction for all of the electrons ✓, and making them more difficult to remove ✓.

> A very good and complete answer with three distinct and relevant points made.

(b) In the third element, an electron has to go into a p-orbital ✓. These are on average further from the nucleus and the electrons are not held so tightly as electrons in s-orbitals. This makes them easier to remove ✓.

> Good! The candidate recognises that electrons in p-orbitals are, on average, further from the nucleus and so it requires less energy to remove them.

(c) Oxygen has a pair of electrons in the 2p-orbital. These repel one another making it easier to remove one of them ✓. In nitrogen, there is one electron in each 2p-orbital ✓.

> The candidate makes two distinct, correct points, one explaining the situation in oxygen, and one relating this to the electron arrangement in nitrogen.

(d) (i) Potassium has an extra shell of electrons between the outer electron and the nucleus making it easier to remove the electron ✓.

> This is correct, no further explanation is needed.

(ii) When they react with water, sodium fizzes and dissolves ✓, but potassium burns with a lilac-pink flame ✗.

> This is a pity. If the candidate had mentioned that it is the hydrogen that burns there would have been two correct and linked observations here.

(e) These elements hadn't been discovered when Mendeleev wrote his table ✓.

> This is correct.

Question 3

Ethanol is considered as an important replacement fuel for petrol in a number of countries.

(a) When ethanol is used as a fuel, combustion takes place as shown in the equation below. The table shows values for standard enthalpy changes of formation, $\Delta H_f^\ominus$.

$$C_2H_5OH(l) + 3O_2(g) \rightarrow 2CO_2(g) + 3H_2O(l) \quad \Delta H^\ominus = -1367\,kJ\,mol^{-1}$$

Compound	$\Delta H_f^{\ominus}$/kJ mol^{-1}
CO_2(g)	–394
H_2O(l)	–286

(i) Define the term *standard enthalpy change of formation.* (3 marks)

(ii) Calculate the standard enthalpy change of formation of ethanol. (3 marks)

(iii) On the axes provided, sketch the enthalpy profile diagram for the combustion of ethanol. Label $\Delta H^{\ominus}$ and E_a on your sketch. (3 marks)

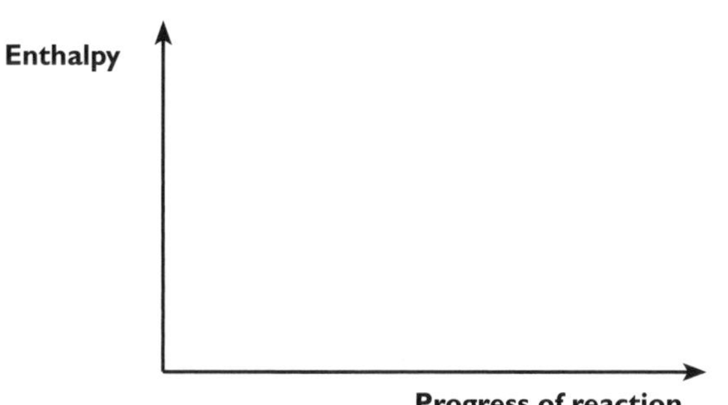

(b) Ethanol can be produced from ethene by reacting it with steam in the presence of a catalyst. The reaction is reversible.

$$C_2H_4(g) + H_2O(g) \rightleftharpoons C_2H_5OH(g)$$

The table shows the percentage conversion of ethene using excess steam under different conditions.

Pressure/atm	Temperature/°C	Conversion/%
40	200	37
40	300	25
75	200	55
75	300	40

(i) Explain the effect of increasing the pressure on the percentage conversion. (2 marks)

(ii) Deduce the sign of the enthalpy change for this reaction, explaining how you arrived at your conclusion. (2 marks)

(iii) The equation shown for the reaction shows one mole of ethene reacting with one mole of steam. Why is excess steam used in the industrial process? (1 mark)

Total: 14 marks

Candidate A

(a) (i) This is the enthalpy change when 1 mol of substance ✓ is formed from its elements ✓.

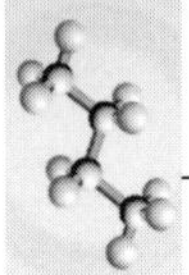

This definition is not quite complete. The candidate has omitted to mention that it has to be under standard conditions.

(ii) To calculate this, I need to reverse the equation and add the equations for the formation of CO_2 and H_2O ✗.

+1367 – 394 – 286 = 687 kJ mol^{-1} ✓ ✓ (ecf)

This is an unfortunate mistake. The candidate has forgotten that the equations for the formation of CO_2 and H_2O need to be multiplied by 2 and 3 respectively to match the combustion of ethanol equation when reversed. The maths based on this faulty logic is, however, correct and so scores 'ecf' (error carried forward) marks.

(iii)

Enthaply

E_a

Reactants

$\Delta H^{\ominus}$

Products

Progress of reaction

Shape ✓, E_a ✓, $\Delta H^{\ominus}$ ✗

This is a good attempt, with the shape and E_a both shown correctly.

(b) (i) Higher pressure means better conversion ✓.

While this is a correct statement about the effect of higher pressure, it does not explain why it occurs, so only scores 1 mark.

(ii) It's an exothermic reaction.

The candidate has made a correct statement but does not gain any marks. This is because the question has not been answered. This is an easy and costly error to make. In each examination many candidates lose marks by not reading the question, or by giving an answer to a slightly different question.

(iii) It pushes the equilibrium to the right ✓.

This question can be answered in a number of different but equally acceptable ways. This is perfectly correct.

Candidate B

(a) (i) This is the enthalpy change when 1 mol of substance ✓ is formed from its elements ✓ under standard conditions ✓.

This candidate has learnt the definition thoroughly and given a complete answer.

(ii) Reverse the equation for the formation of ethanol, then add 2 × $\Delta H^\ominus_f$ for CO_2 and 3 × $\Delta H^\ominus_f$ for H_2O ✓.

+1367 − 2 × 394 − 3 × 286 = −279 kJ mol^{-1} ✓ ✓

In this calculation, the candidate explains what is being done and why. As a result, the calculation of $\Delta H^\ominus_f$ for ethanol is correct.

(iii)

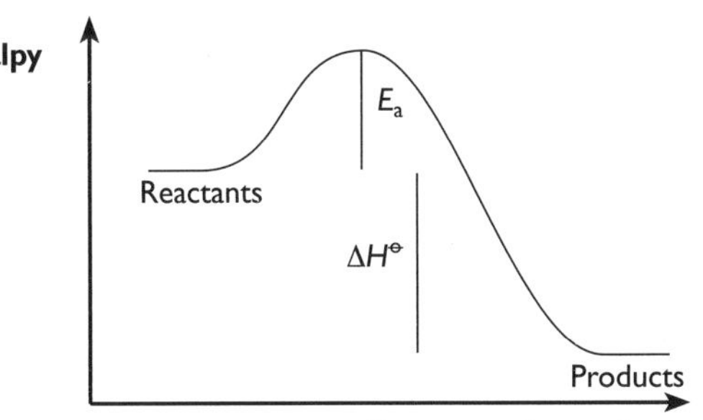

Shape ✓, E_a ✓, $\Delta H^\ominus$ ✓

Although the candidate gains all 3 marks, the answer is not quite perfect. It would have been better to show arrowheads on the lines for E_a and $\Delta H^\ominus$.

(b) (i) A higher percentage of ethene is converted at higher pressure ✓ because there are fewer gas particles on the right-hand side of the equilibrium ✗.

The first statement here is true, but the second statement does not go quite far enough. There are fewer gas particles on the right-hand side of the equilibrium, but the point is that this shifts the equilibrium to the right, yielding more ethanol.

(ii) The forward reaction is exothermic so the sign is negative ✓. Less ethene is converted at higher temperatures because the equilibrium moves to the left to try to reduce the temperature ✓.

A good, full answer showing that this candidate understands Le Chatelier's principle. However, it is a pity that the answer to part (i) did not include this detail.

(iii) To make sure as much ethene as possible reacts ✓.

This is a different answer from that given by Candidate A, but it is still correct.

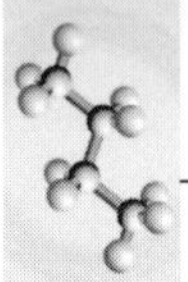

Question 4

Group VII is the only group in the periodic table that has elements that are in gaseous, liquid and solid states at room temperature. In this question you need to consider the three elements, chlorine, bromine and iodine.

(a) All three elements exist as diatomic molecules with the atoms linked by a single covalent bond. Explain the differences in their volatility. (2 marks)

(b) The table shows the bond energies for the three hydrogen halides.

Bond energy / kJ mol^{-1}		
HC*l*	HBr	HI
432	366	298

(i) **Explain the trend in bond energies shown.** (2 marks)

(ii) **Describe the effect of plunging a red-hot wire into tubes containing each of the hydrogen halides.** (2 marks)

(c) Chlorine reacts with hot aqueous sodium hydroxide according to the following equation:

$$3Cl_2(aq) + 6OH^-(aq) \rightarrow 5Cl^-(aq) + ClO_3^-(aq) + 3H_2O(l)$$

State the oxidation number of chlorine in each chlorine-containing compound formed. (2 marks)

Total: 8 marks

Candidate A

(a) Down the group the charge on the nucleus increases.

No marks can be awarded here because, although this is a true statement, it does not answer the question. Beware of just writing 'snippets' of chemistry from this area of the syllabus. Think what you need to say in order to answer the question.

(b) (i) As the halogen atoms get bigger ✓, the overlap between orbitals with hydrogen gets less. This weakens the H–X bond ✓.

This good answer is complete and to the point.

(ii) HC*l* no effect, HBr some decomposition, HI lots of decomposition ✓.

Although the candidate has indicated the correct consequences, for 1 mark, there is no real description of what would be observed.

(c) –1 in Cl^- ✓ and +5 in ClO_3^- ✓

In this correct answer, the candidate does not just state the oxidation number, but relates it to the correct ion.

Candidate B

(a) The molecules are held together by van der Waals forces ✓.

The larger the atoms, the greater the van der Waals forces and hence the higher the boiling point ✓.

This is a very good descriptive answer that links the type of force/bonding involved with the reason for the increasing boiling points.

(b) (i) As we descend the group the halogen atoms increase in size ✓. This reduces the overlap between orbitals on the hydrogen atom and the halogen atom, weakening the H–X bond ✓.

This is another good and full answer. It is expressed differently from the answer given by Candidate A, but both answers are correct.

(ii) H*Cl* no effect, HBr some decomposition with a yellow-orange tinge, HI lots of decomposition with violet vapour ✓ ✓.

Here the candidate describes both the degree of reaction and what would be observed to support this trend.

(c) 1 in Cl^- ✗ and +5 in ClO_3^- ✓

This is a pity. The minus sign is missing from in front of the 1 for Cl^- making the answer incorrect. The response for ClO_3^- is correct.

Question 5

Two of the forms of isomerism found in organic compounds are structural isomerism and *cis–trans* isomerism.

(a) Two of the structural isomers of $C_3H_4Cl_2$ are shown below.

Isomer 1: $(H)(H_3C)C{=}C(Cl)(Cl)$

Isomer 2: $(H)(CH_2Cl)C{=}C(Cl)(H)$

(i) Draw the two other structural isomers of $C_3H_4Cl_2$ that are chloroalkenes and that contain a methyl group. (2 marks)

(ii) Draw another structural isomer of $C_3H_4Cl_2$ that is *not* a chloroalkene. (1 mark)

(iii) How many isomers of $C_3H_4Cl_2$ in total contain the $-CH_2Cl$ group? (1 mark)

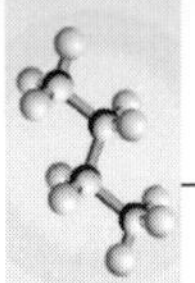

(b) (i) Which, if any, of the structural isomers, including isomers 1 and 2 and those that you have drawn, are *cis–trans* isomers? (1 mark)

(ii) Draw the two possible products you could get if the *cis*-isomer was reacted with HBr. (2 marks)

Total: 7 marks

Candidate A

(a) (i)

H_3C, Cl / C=C / H, Cl ✗
Isomer 3

Cl, Cl / C=C / CH_3, H ✓
Isomer 4

Isomer 3 is the same as isomer 1, but rotated 180° around the double bond.

(ii)

Cl, Cl — C — CH_2 — H_2C (cyclic) ✓

This is correct. The candidate did well to spot that this has to be a cyclic compound if it isn't an alkene!

(iii) Two ✗

This candidate is not sure about isomers of compounds that contain a double bond. The candidate has also not been logical since both isomers 3 and 4 contain a methyl group, as does isomer 1.

(b) (i) 3 and 4 ✗

The candidate once more shows confusion about *cis–trans* isomerism. For this to exist, the atoms or groups have to be at opposite ends of the double bond.

(ii) $CH_3CClBr–CH_2Cl$ ✓ and $CH_3CHCl–CHClBr$ ✓

Apparently the candidate does know what *cis*- means! These are good answers, but would have been clearer if the formulae had been displayed.

Candidate B

(a) (i)

Cl, Cl / C=C / CH_3, H ✓
Isomer 4

Cl, H / C=C / CH_3, Cl ✓
Isomer 5

Both isomers are correct.

(ii)

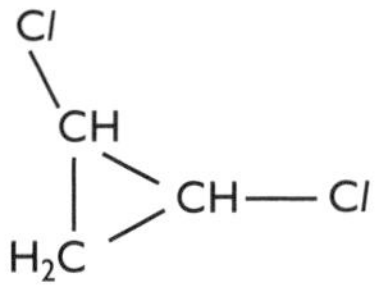

```
 Cl
  \
   CH
   | \
   |  CH—Cl
   | /
  H2C
```
✓

This is a different cyclic isomer from that given by Candidate A, but is still correct.

(iii) Three ✓

This leads from a sound understanding of isomerism and the correct answer in (a)(i).

(b) (i) 4 and 5 ✓

Good! This is once more made simple after a correct answer to (a)(i). This candidate's answers illustrate how a good grasp of the basic principles can lead to much better marks.

(ii)

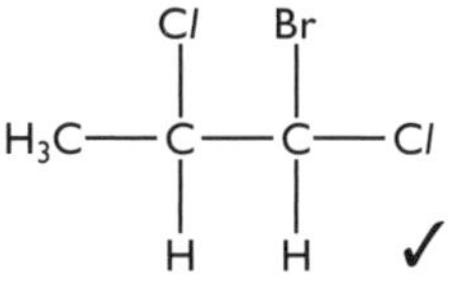

```
        Cl   Br                    Cl   H
        |    |                     |    |
H3C—C——C—Cl       and     H3C—C——C—Cl
        |    |                     |    |
        H    H   ✓                 Br   H   ✓
```

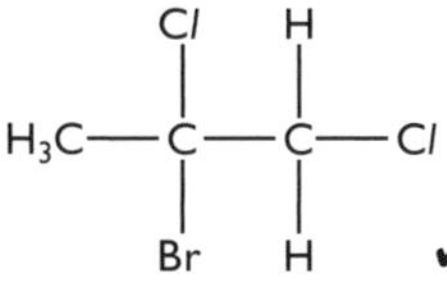

This is a well thought out answer.

Question 6

The diagram shows some of the reactions of menthol, a naturally-occurring alcohol found in peppermint oil.

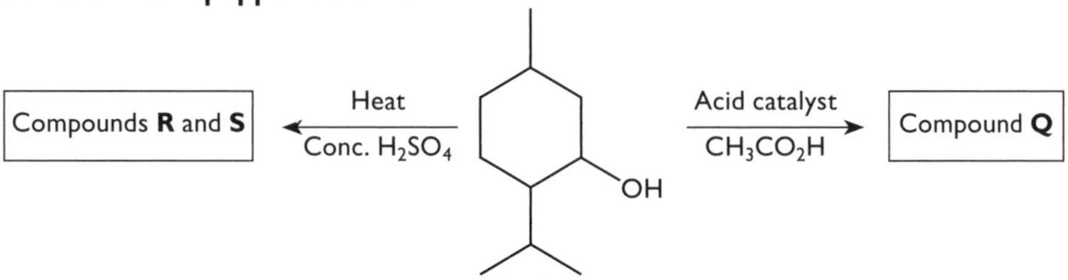

Compounds **R** and **S** ← (Heat, Conc. H_2SO_4) — Menthol (OH) — (Acid catalyst, CH_3CO_2H) → Compound **Q**

Menthol

(a) (i) The structure of menthol is shown in skeletal form. What is its molecular formula? (1 mark)

(ii) Is menthol a primary, secondary or tertiary alcohol? (1 mark)

(b) (i) What type of reaction forms compound Q? (1 mark)

(ii) Compounds R and S have the same molecular formula. Draw their structures. (2 marks)

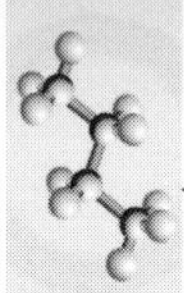

(iii) Compound R is treated with aqueous bromine. State what you would observe. (1 mark)

(c) Menthol can be oxidised to form compound T, which forms an orange-yellow precipitate with 2,4-dinitrophenylhydrazine.

(i) Give the reagents and conditions for the formation of T. (2 marks)

(ii) What would you observe during the oxidation? (1 mark)

Total: 9 marks

Candidate A

(a) (i) $C_{10}H_{17}O$ ✗

The candidate seems to have assumed that the cyclohexane ring is a benzene ring and has miscalculated the number of hydrogen atoms.

(ii) Secondary ✓

This is correct.

(b) (i) Making an ester ✓

Good, the candidate recognises that the reaction is between an organic acid and an alcohol.

(ii)

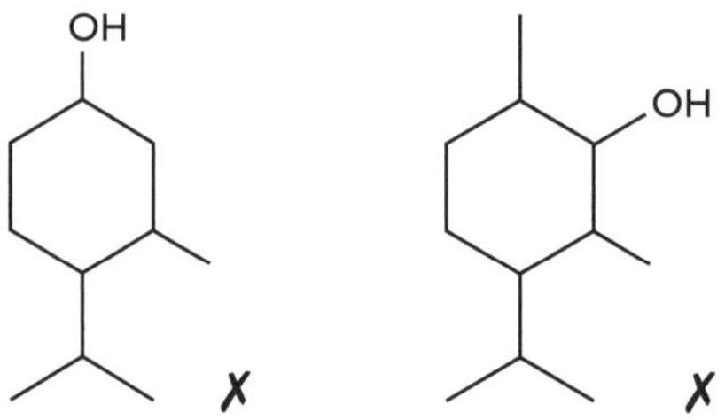

OH ✗ OH ✗

There is confusion here. The candidate seems to think that the –OH group is not removed but can migrate to different ring positions.

(iii) The bromine would turn colourless ✓

This is not a perfect answer, but it is close enough to earn the mark.

(c) (i) Heat ✓ with potassium manganate(VII) ✗

The candidate uses the correct substance but to earn both marks acidified aqueous managanate(VII) should be specified.

(ii) The purple solution would turn pink ✗.

This is not sufficiently accurate. The correct answer is 'very pale pink' or 'colourless'.

Candidate B

(a) (i) $C_{10}H_{20}O$ ✓

This is correct.

(ii) Secondary ✓

This is correct.

(b) (i) Esterification ✓

Good, this candidate also recognises that the reaction is between an organic acid and with an alcohol to form an ester.

(ii)

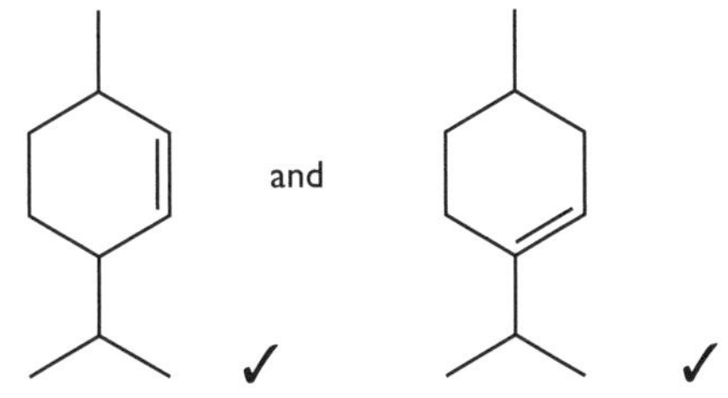

This is a very good answer. Not only has the candidate worked out that an elimination reaction takes place giving a double bond in the ring, but that it could involve one of two different hydrogen atoms and the –OH group.

(iii) The orange colour of the bromine would disappear ✓.

This is a full and correct answer. Compare it with that of Candidate A.

(c) (i) Heat with acidified ✓ potassium dichromate(VI) solution ✓.

Both the correct reagent and conditions are given, for 2 marks.

(ii) The dichromate solution would be decolorised ✗.

To earn the mark, the initial colour and the change should both be given. The orange dichromate solution is not decolorised — it turns green.

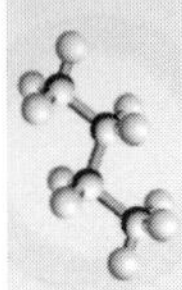

A2 exemplar paper

Question 1

(a) Define the term *standard electrode potential.* (3 marks)

(b) Draw a labelled diagram to show how you would measure the electrode potential of Pb/Pb^{2+}. (4 marks)

(c) Using relevant $E^{\ominus}$ data from the *Data Booklet*, explain how these data relate to the relative reactivity of chlorine, bromine and iodine as oxidising agents. (2 marks)

(d) Use relevant $E^{\ominus}$ data from the *Data Booklet* to construct redox equations, and calculate the standard cell potentials for the reactions between:

(i) $SO_2(aq)$ and $Br_2(aq)$

(ii) acidified $H_2O_2(aq)$ and $MnO_4^-(aq)$ (4 marks)

Total: 13 marks

Candidate A

(a) The standard electrode potential is the potential of the electrode measured under standard conditions ✓ using molar solutions ✓.

This definition is almost correct, for 2 marks. It does not mention that the potential is measured against a standard hydrogen electrode, so the third mark is lost.

(b)

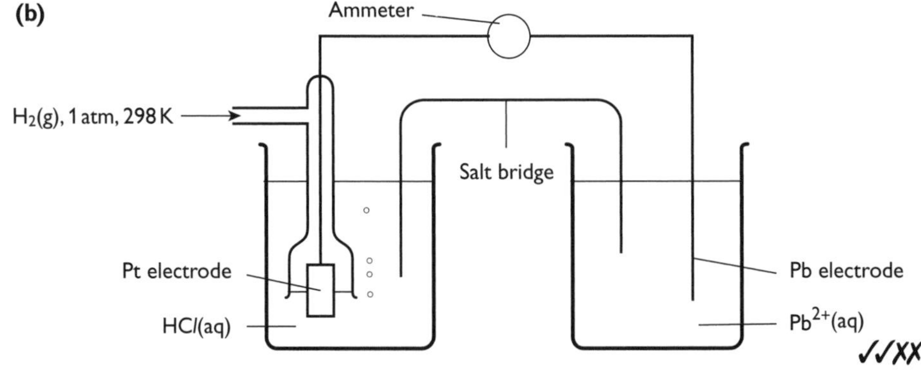

This is a nice clear diagram with just two mistakes — the ammeter should be a voltmeter, and there is no concentration (1.0 mol dm^{-3}) given for the solutions.

(c) The *Data Booklet* gives the $E^{\ominus}$ data for chlorine, bromine and iodine as +1.36 V, +1.07 V and +0.54 V respectively. The positive sign shows that the elements are oxidising agents ✓. The larger the value of $E^{\ominus}$ the better the oxidising agent ✓.

This is a well-reasoned answer.

(d) (i) The two half-equations are:

$SO_2 + 2H_2O \rightarrow SO_4^{2-} + 4H^+ + 2e^-$ $E^\ominus = -0.17\,V$
$Br_2 + 2e^- \rightarrow 2Br^-$ $E^\ominus = +1.07\,V$

To work out the equation I need to add these:

$SO_2 + 2H_2O + Br_2 \rightarrow SO_4^{2-} + 4H^+ + 2Br^-$ ✓

$E^\ominus_{cell} = -0.17 + 1.07 = +0.90\,V$ ✓

This is a good, well laid out and easy to follow answer.

(ii) The two half-equations are:

$MnO_4^- + 8H^+ + 5e^- \rightarrow Mn^{2+} + 4H_2O$ $E^\ominus = +1.52\,V$
$H_2O_2 + 2H^+ + 2e^- \rightarrow 2H_2O$ $E^\ominus = +1.77\,V$ ✗

To work out the equation I need to reverse the second equation and multiply by 5, then add this to twice the first equation:

$2MnO_4^- + 16H^+ + 10e^- \rightarrow 2Mn^{2+} + 8H_2O$
$10H_2O \rightarrow 5H_2O_2 + 10H^+ + 10e^-$

$2MnO_4^- + 16H^+ + 2H_2O \rightarrow 2Mn^{2+} + 5H_2O_2$

$E^\ominus_{cell} = 1.52 - 1.77 = -0.25\,V$ ✓ (ecf)

The answers are again well laid out and easy to follow, but the candidate has chosen an incorrect half-equation for hydrogen peroxide, where hydrogen peroxide is reduced, not oxidised. The value for $E^\ominus_{cell}$ has been calculated correctly, so there is an 'error carried forward' mark here.

Candidate B

(a) The standard electrode potential is the potential of the electrode measured under standard conditions ✓ against a standard hydrogen electrode ✓.

This definition is almost correct, but does not mention that the solutions need to be molar. Perhaps the candidate assumed this under 'standard conditions' because it is shown in the diagram in part (b). However, you must not assume that the examiner will understand what is in your mind!

(b)

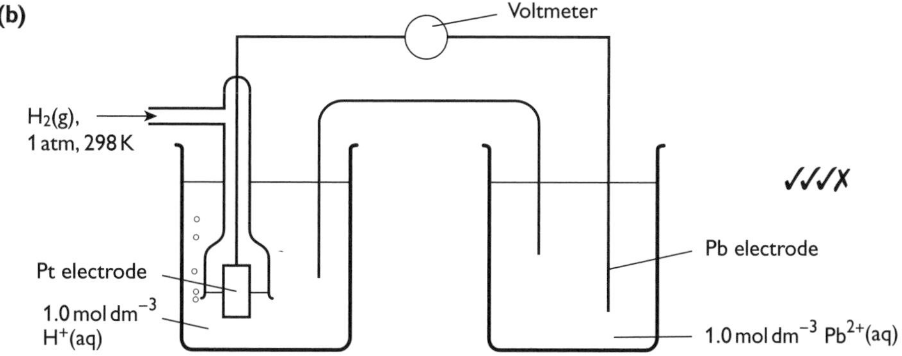

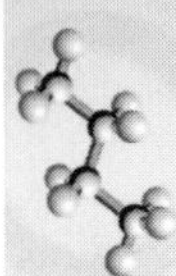

This is a nice clear diagram with just one mistake — the salt bridge is not labelled.

(c) The *Data Booklet* shows that the $E^\ominus$ data for chlorine, bromine and iodine are all positive, indicating that they are oxidising agents ✓. The values +1.36 V, +1.07 V and +0.54 V respectively tell us that the larger the value of $E^\ominus$ the stronger the oxidising agent ✓.

This is a well-reasoned answer that uses slightly more precise language than Candidate A.

(d) (i) The two half-equations are:

$SO_2 + 2H_2O \rightarrow SO_4^{2-} + 4H^+ + 2e^-$ $E^\ominus = -0.17\,V$
$Br_2 + 2e^- \rightarrow 2Br^-$ $E^\ominus = +1.07\,V$

Since the electrons balance, to work out the equation I need to just add the half-equations:

$SO_2 + 2H_2O + Br_2 \rightarrow SO_4^{2-} + 4H^+ + 2Br^-$ ✓

$E^\ominus_{cell} = -0.17 + 1.07 = +0.90\,V$ ✓

This is a good answer that is well laid out and easy to follow.

(ii) The two half-equations are:

$MnO_4^- + 8H^+ + 5e^- \rightarrow Mn^{2+} + 4H_2O$ $E^\ominus = +1.52\,V$
$O_2 + 2H^+ + 2e^- \rightarrow H_2O_2$ $E^\ominus = +0.68\,V$

To make the electrons balance and hence work out the equation I need to reverse the second equation and multiply it by 5, and double the first equation. Then add the two equations together.

$2MnO_4^- + 16H^+ + 10e^- \rightarrow 2Mn^{2+} + 8H_2O$
$5H_2O_2 \rightarrow 5O_2 + 10H^+ + 10e^-$

$2MnO_4^- + 16H^+ + 5H_2O_2 \rightarrow 2Mn^{2+} + 5O_2 + 8H_2O$

$E^\ominus_{cell} = 1.52 - 0.68 = +0.84\,V$ ✓

This is very good. The answers are again well laid out and easy to follow, even though this is a harder reaction to work with.

Question 2

Both silver chloride and silver bromide have been used in photographic film for many years. The compounds form a thin emulsion on the film and when exposed to light, the halide ion absorbs a photon and releases an electron. This electron then reduces the silver ion to metallic silver.

$\mathbf{Br^- + h\nu \rightarrow Br + e^-}$

$\mathbf{Ag^+ + e^- \rightarrow Ag}$

(a) Predict whether it would need more energy or less energy for this process for AgC*l* than AgBr. Explain your answer. (1 mark)

(b) Write a chemical equation to represent the lattice energy of AgC*l*. (1 mark)

(c) Using the following data, calculate the lattice energy of AgC*l*(s).

First ionisation energy of silver	$+731\,kJ\,mol^{-1}$
Electron affinity of chlorine	$-349\,kJ\,mol^{-1}$
Enthalpy change of atomisation of silver	$+285\,kJ\,mol^{-1}$
Enthalpy change of atomisation of chlorine	$+121\,kJ\,mol^{-1}$
Enthalpy change of formation of AgC*l*(s)	$-127\,kJ\,mol^{-1}$

(3 marks)

(d) How might the lattice energy of AgBr(s) compare with that of AgC*l*(s)? Explain your answer. (2 marks)

(e) Silver chloride is not very soluble in water. This makes the formation of a white precipitate on addition of silver ions to a solution containing chloride ions a useful analytical test. The solubility of silver chloride is $1.3 \times 10^{-5}\,mol\,dm^{-3}$.

(i) Write an expression for the solubility product of silver chloride. (1 mark)

(ii) Calculate the value of the solubility product and give its units. (2 marks)

(f) If ammonia solution is added to a freshly formed precipitate of silver chloride, the precipitate dissolves to give a colourless solution. Write a balanced equation for this reaction. (2 marks)

Total: 12 marks

Candidate A

(a) Chlorine is more reactive than bromine so will be harder to change from Cl^- to Cl ✓.

The logic here is fine, even if it is not precise.

(b) $Ag^+(g) + Cl^-(g) \rightarrow AgCl(s)$ ✓

Although the AgC*l* is ionic, charges do not have to be shown since it is a solid.

(c) To find the lattice energy use a Born–Haber cycle:

$+285 + 121 + 731 + 349$ ✗ $+ \Delta H^{\circ}_{L} = -127$

$\Delta H^{\circ}_{L} = -127 - 285 - 121 - 731 - 349$ ✓

$= -1613\,kJ\,mol^{-1}$ ✓ (ecf)

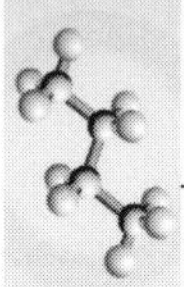

There is one error in the Born–Haber calculation — the sign of the electron affinity of chlorine has been copied incorrectly. The remainder of the calculation is correct scoring two 'error carried forward' marks.

(d) The value of the lattice energy of AgBr differs from that of Ag*Cl* because of two factors — the enthalpy change of atomisation and electron affinity of bromine and chlorine are different. For bromine the energy change of atomisation is less positive; the electron affinity is less negative. Together these make the lattice energy less positive because the electron affinity has the greater difference ✓.

This is a good try for 1 mark. However, there is another term that affects the value of the lattice energy. This is the enthalpy of formation of silver bromide compared with that of silver chloride.

(e) (i) $K_{sp} = [Ag^+(aq)][Cl^-(aq)]$ ✓

This is correct.

(ii) $K_{sp}(AgCl) = 1.3 \times 10^{-5} \times 1.3 \times 10^{-5} = 1.69 \times 10^{-10}\,mol\,dm^{-3}$ ✓ ✗

The calculation is correct but the units are wrong.

(f) $Ag^+Cl^- + 2NH_3 \rightleftharpoons [Ag(NH_3)_2]^+ + Cl^-$ ✓

This is correct.

Candidate B

(a) Bromine forms Br^- less easily than chlorine forms Cl^-, so the reverse reaction should be easier ✓.

This is correct.

(b) $Ag^+(g) + Cl^-(g) \rightarrow Ag^+(s)Cl^-(s)$ ✗

This is incorrect. It is the Ag*Cl* that is a solid, not the individual ions.

(c) To find the lattice energy I need to use a Born–Haber cycle.

$E_a(Ag) + E_a(Cl) + 1st\ IE(Ag) + EA(Cl) + LE(AgCl) = E_f(AgCl)$

$+285 + 121 + 731 - 349 + \Delta H^{\circ}_L = -127$

$\Delta H^{\circ}_L = -127 - 285 - 121 - 731 + 349$

$= -915\,kJ\,mol^{-1}$ ✓ ✓ ✓

This is completely correct. The answer is well laid-out and easy for an examiner to follow.

(d) When we consider AgBr rather than Ag*Cl*, three values in the Born–Haber cycle change — $E_a(Cl)$, $EA(Cl)$ and $E_f(AgCl)$. The electron affinity of bromine is less negative than that of chlorine, the energy of atomisation of bromine is less positive than that of chlorine, and the enthalpy of formation of silver bromide will be less negative than that of silver chloride. Overall this suggests that the lattice energy of silver bromide will be less negative ✓ ✓.

This very good complete answer scores both marks.

(e) (i) $K_{sp} = [Ag^+]\ [Cl^-]$ ✓

This is correct.

(ii) $K_{sp}(AgCl) = 1.3 \times 10^{-5} \times 1.3 \times 10^{-5} = 1.69 \times 10^{-10}\,mol^2\,dm^{-6}$ ✓ ✓

Both the calculation and the units are correct.

(f) $Ag^+Cl^- + 2NH_3 \rightleftharpoons [Ag(NH_3)_2]^+ + Cl^-$ ✓

This is correct.

Question 3

(a) Nitric acid reacts with bases such as aqueous ammonia, $NH_3(aq)$, to form salts. A 25.0 cm³ sample of nitric acid was pipetted into a conical flask. Aqueous ammonia was added from a burette until little further change in pH of the solution was observed. The resulting pH curve for the titration is shown below.

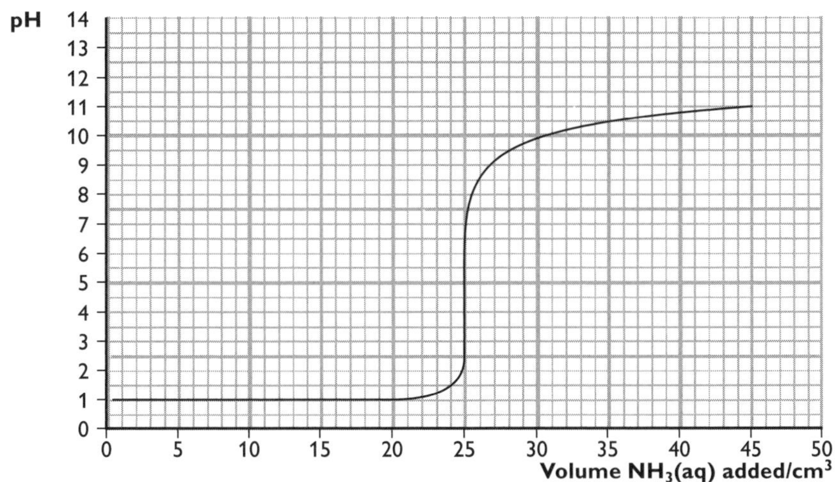

(i) Estimate the pH at the end point of this titration. (1 mark)

(ii) How can you tell from the pH curve that aqueous ammonia is a weak base? (1 mark)

(iii) What was the concentration of the nitric acid? (1 mark)

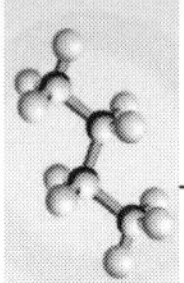

The pH ranges of four indicators are given in the table.

Indicator	pH range
Gentian violet	0.1–2.0
Methyl red	4.4–6.2
Cresol red	7.2–8.8
Alizarin yellow R	10.1–12.0

(iv) Explain which of these indicators would be most suitable to use for this titration. (1 mark)

(b) A buffer solution was made by mixing 50 cm^3 of 0.100 mol dm^{-3} sodium ethanoate solution with 50 cm^3 of 0.100 mol dm^{-3} ethanoic acid solution. K_a for ethanoic acid is 1.8 × 10^{-5} mol dm^{-3}. Calculate the pH of the buffer solution. Show your working. (2 marks)

(c) The equation shows the dissociation of ethanoic acid:

$$CH_3CO_2H(aq) + H_2O(l) \rightleftharpoons CH_3CO_2^-(aq) + H_3O^+(aq)$$

Explain the effect of adding the following to this solution:

(i) H_3O^+ ions (1 mark)

(ii) OH^- ions (1 mark)

Total: 8 marks

Candidate A

(a) (i) pH 5 ✓

This is correct.

(ii) The curve ends below pH 14 ✓.

Although this is true, it does not provide evidence that ammonia is a weak base.

(iii) 1.0 mol dm^{-3} ✗

This is incorrect. If the pH = 1, then $[H^+] = 1 \times 10^{-1} = 0.1$ mol dm^{-3}

(iv) Methyl red, because it changes colour during the vertical portion of the curve ✓.

This is a good answer.

(b) $pH = -\log(1.8 \times 10^{-5}) + \log\frac{(0.100)}{(0.100)}$ ✗

$= -\log(1.8 \times 10^{-5}) + \log 1$

$= -\log(1.8 \times 10^{-5})$ because $\log 1 = 0$

$= 4.7$ ✓

This is perhaps a little harsh. However, the concentrations of the two solutions need to be halved to 0.050 mol dm^{-3} — they happen to cancel out here.

(c) (i) On adding H_3O^+ ions these will combine with $CH_3CO_2^-$ ions shifting the equilibrium to the left ✓.

This is a good answer.

(ii) On adding OH^- ions these will react with H_3O^+ ions shifting the equilibrium to the right ✓.

This is another good answer.

Candidate B

(a) (i) pH 5 ✓

This is correct.

(ii) The curve does not reach pH 14. It flattens out around pH 11.5 ✓.

A correct statement based on the data from the titration curve.

(iii) Since the starting pH is 1, $-\log[H^+] = 1$ and hence the concentration of the acid must be $1 \times 10^{-1} = 0.1\,mol\,dm^{-3}$ ✓.

This is correct and well explained.

(iv) Methyl red because the mid-point of its pH range is close to the mid-point of the vertical portion of the curve ✓.

This is another correct, well explained answer.

(b) $pH = -\log(1.8 \times 10^{-5}) + \log\frac{(0.050)}{(0.050)}$

$= -\log(1.8 \times 10^{-5}) + \log 1$

$= -\log(1.8 \times 10^{-5})$ because $\log 1 = 0$

$= 4.7$ ✓ ✓

Both the calculation and the units are correct.

(c) (i) When H_3O^+ ions are added, they will combine with $CH_3CO_2^-$ ions, thus shifting the equilibrium to the left ✓.

This is a good answer.

(ii) When OH^- ions are added, they will react with H_3O^+, thus shifting the equilibrium to the right ✓.

This is another good answer.

Question 4

The metals of Group II and the transition metals show very different properties.

(a) Give *three* examples of differences in *chemical* properties between these two groups of metals. (3 marks)

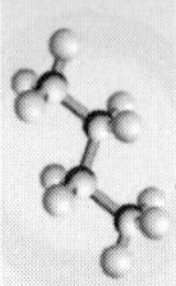

(b) Anhydrous copper(II) ions, Cu^{2+}, are colourless. Aqueous copper(II) ions, $[Cu(H_2O)_6]^{2+}$, are pale blue. Copper(II) ions complexed with ammonia, $[Cu(NH_3)_4(H_2O)_2]^{2+}$, are deep blue-purple. Explain these observations in terms of your knowledge of copper(II) and its d-orbitals. (3 marks)

(c) Copy the axes below. Sketch the visible spectrum of A, $[Cu(H_2O)_6]^{2+}$, and B, $[Cu(NH_3)_4(H_2O)_2]^{2+}$, on the axes, labelling clearly which is which. (2 marks)

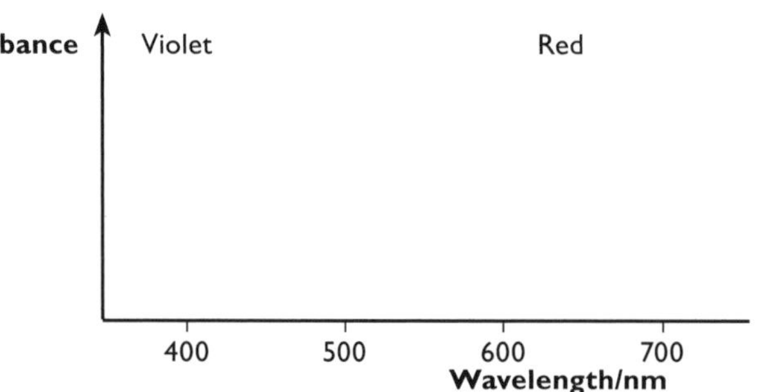

(d) Which of the d-orbitals are used in forming octahedral complexes such as $[Cu(H_2O)_6]^{2+}$? (1 mark)

(e) When the manganate(VII) ion, MnO_4^-, reacts with sulfur dioxide, SO_2, in acid solution, the deep purple solution turns almost colourless.

(i) Write a balanced equation for this reaction. (2 marks)

(ii) Suggest a reason why the new solution has little colour. (1 mark)

Total: 12 marks

Candidate A

(a) Transition metals form coloured compounds ✓, they behave as catalysts ✓ and they have high melting points ✗.

> The first two properties are chemical in nature, but the third is physical.

(b) Colours are due to splitting of the d-orbitals into groups of higher and lower energy because of ligands. In anhydrous copper(II) no ligands are present, so there is no colour ✓. In $[Cu(H_2O)_6]^{2+}$ the complex emits ✗ blue because there is a small energy gap between the two sets of orbitals. In $[Cu(NH_3)_4(H_2O)_2]^{2+}$ the gap is bigger, so it emits violet ✓. (ecf)

> The candidate points out correctly that colour in transition metal compounds is a result of the d-orbitals splitting into two groups of different energy. The mistake made is in believing that energy is emitted when electrons move from high to low energy, rather than energy being absorbed in promoting electrons. Hence the second mark is lost. The third mark has been awarded consequentially.

(c)

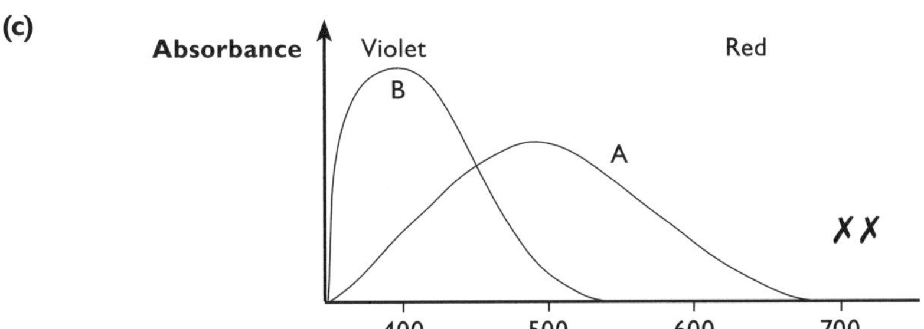

The candidate has drawn two emission spectra rather than absorption spectra.

(d) $3d_{x^2-y^2}$ and $3d_{z^2}$ ✓

This is correct.

(e) (i) $2MnO_4^- + 3SO_2 + 2H_2O \rightarrow 2MnO_2 + 3SO_4^{2-} + 4H^+$ ✗ ✓ (ecf)

The candidate has used a wrong half-equation for manganate(VII). However, the overall equation is balanced correctly and so scores the second mark as an 'error carried forward'.

(ii) The manganese is all precipitated as MnO_2 ✗.

This is an unfortunate consequence of the incorrect equation. No mark can be awarded here because the candidate has ignored data given in the question about the colour change.

Candidate B

(a) Transition metals behave as catalysts ✓, they form complexes with ligands ✓ and many of their compounds are coloured in aqueous solution ✓.

Three correct differences in chemical properties are given.

(b) Colours are due to splitting of the d-orbitals into groups of higher and lower energy because of ligands. In anhydrous copper(II) there are no ligands, so there is no colour ✓. In $[Cu(H_2O)_6]^{2+}$ the complex absorbs orange and red light because there is a small energy gap between the two sets of orbitals ✓. In $[Cu(NH_3)_4(H_2O)_2]^{2+}$ the gap is bigger, so it transmits violet ✗.

The candidate points out correctly that colour in transition metal compounds is a result of the d-orbitals splitting into two groups of different energy and that the complex $[Cu(H_2O)_6]^{2+}$ absorbs orange and red light (and by implication transmits the rest). The statement 'in $[Cu(NH_3)_4(H_2O)_2]^{2+}$ the gap is bigger' is correct, but the third mark is lost because of failure to refer to the colour of light absorbed.

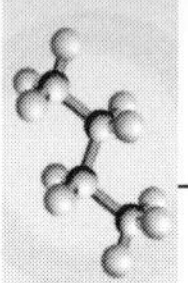

(c)

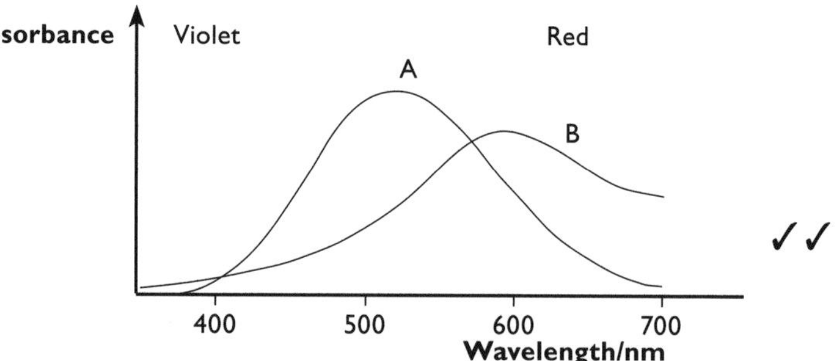

✓✓

The candidate has correctly drawn two absorption spectra. In questions like this the examiners are looking for correct principles being illustrated rather than perfect representations.

(d) $3d_{x^2-y^2}$ and $3d_{z^2}$ ✓

This is correct.

(e) (i) $2MnO_4^- + 5SO_2 + 2H_2O \rightarrow 2Mn^{2+} + 5SO_4^{2-} + 4H^+$ ✓ ✓

This is the correct equation, correctly balanced.

(ii) Mn^{2+} has one electron in each of its 3d-orbitals. To promote an electron would require energy to overcome the repulsion of putting two electrons in the same orbital. This does not occur readily, making Mn^{2+} almost colourless ✓.

This is a good answer. The candidate recognises the critical point that the manganese is present as Mn^{2+} at the end of the reaction.

Question 5

The common analgesic drug paracetamol can be made from phenol in three steps.

phenol (OH) —1→ **P** (OH, NO_2) —2→ **Q** (OH, NH_2) —3→ **Paracetamol** (OH, $NHCOCH_3$)

(a) (i) Suggest the reagents and conditions for step 1. (2 marks)

(ii) What type of reaction is step 2? (1 mark)

(iii) What reagents and conditions would you use for step 2? (2 marks)

(iv) Name a reagent that could be used for step 3. (1 mark)

(v) What *two* functional groups are present in *paracetamol*? (2 marks)

(b) (i) State what you would see if aqueous bromine was added to compound Q. (2 marks)

(ii) Draw the product from the reaction of Q with aqueous HC*l*. (1 mark)

(iii) What sort of reaction is this? (1 mark)

Total: 12 marks

Candidate A

(a) (i) Concentrated nitric and sulfuric acids ✗

The candidate has given the reagents for nitration of benzene. Nitration of phenol takes place under milder conditions (see Candidate B).

(ii) Hydrogenation ✓

Although most textbooks would refer to this as a reduction reaction, hydrogenation would probably be allowed here.

(iii) Tin and concentrated hydrochloric acid, ✓ heat ✓

The correct reagents and conditions are given, for 2 marks.

(iv) CH_3CO_2H ✗

The candidate does not seem to have recognised that an amide has been formed.

(v) Alcohol ✗ and amide ✓

The candidate has forgotten that the –OH group is attached to a benzene ring and is, therefore, a phenol group, not an alcohol.

(b) (i) A white precipitate forms ✓.

The candidate has remembered that a white precipitate is formed with phenols and bromine, but has omitted to mention that the bromine loses its colour as it reacts. Therefore, the candidate scores only 1 of the 2 marks.

(ii)

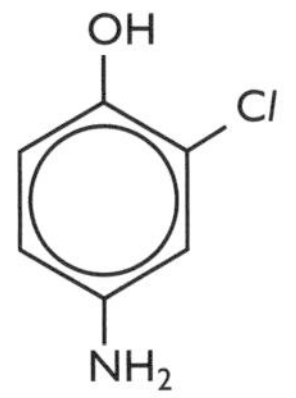

OH

C*l*

NH_2 ✗

This is incorrect. HC*l* does not react to give substitution in the ring.

(iii) Substitution ✓ (ecf)

Although this is an incorrect answer, it describes correctly what the candidate believed to be true in (ii) and so scores an 'error carried forward' mark.

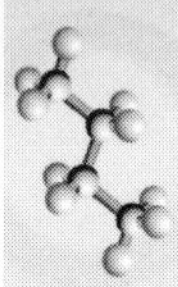

Candidate B

(a) (i) Dilute nitric acid ✓ on warming ✓.

This is a good answer with reagent and conditions correct.

(ii) Reduction ✓

This is correct.

(iii) Lithium aluminium hydride ✓ in ether ✓.

This is another good answer with both reagents and conditions correct.

(iv) Ethanoyl chloride ✓.

This is correct.

(v) Phenol ✓ and amide ✓.

Both groups are correct.

(b) (i) Bromine is decolourised ✓ and a white precipitate is formed ✓.

Both observations are correct.

(ii)

OH

$N^+H_3Cl^-$ ✓

This is correct.

(iii) Neutralisation ✓

Yet another correct answer.

Question 6

Man-made polymers play an important part in the manufacture of a wide range of products. Contact lenses are made using hydrophilic polymers with structures such as the two examples shown below.

L

OH OH OH OH

CH CH CH CH

CH_2 CH_2 CH_2 CH_2

M

OH O

H H_2 H_2

N CH C C C

C CH N CH N

O OH H OH H

(a) What does the term *hydrophilic* mean? (1 mark)

(b) Draw the monomer units for each of the two polymers, L and M. (3 marks)

(c) (i) The polymer chains in L can be cross-linked using a small molecule containing two functional groups. Draw the structure of such a molecule. (1 mark)

(ii) What type of bond would be formed between L and the molecule you have drawn? (1 mark)

Proteins and polypeptides are natural polymers found in living organisms. They are formed by linking different types of amino acid. Two amino acids are shown below.

$H_2NCH_2CO_2H$	$H_2NCH(CH_3)CO_2H$
Glycine (gly)	**Alanine (ala)**

(In alanine, the CH_3 group is attached by a vertical bond to the central C of H_2NCHCO_2H.)

(d) (i) Draw the structure of the dipeptide gly–ala formed from these two amino acids. You should show the peptide bond in displayed form. (2 marks)

(ii) What is unusual about the structure of glycine compared with other amino acids? (1 mark)

(e) A small polypeptide T was broken down into its constituent amino acids with the following outcome:

$$T \rightarrow 2H_2NCH_2CO_2H + 3H_2NCH(CH_3)CO_2H + 2H_2NCH(CH_2CSH)CO_2H$$

(The CH_3 and CH_2CSH groups are shown on vertical bonds below the central C of each H_2NCHCO_2H.)

$M_r = 75$ $M_r = 89$ $M_r = 133$

(i) How many peptide bonds were broken in this reaction? (1 mark)

(ii) Calculate the M_r of T. (1 mark)

(iii) Describe how the polypeptide could be broken down in the laboratory without the use of enzymes. (2 marks)

Total: 13 marks

Candidate A

(a) Liking water ✗.

This is too vague an answer to score the mark.

(b) Monomer of L: $H_2C{=}CHOH$ ✓

Monomers of M: $CO_2HCH(OH)CH(OH)CO_2H$ ✗

$NH_2CH_2CH(OH)CH_2NH_2$ ✓

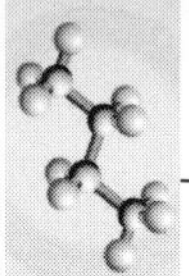

This is quite a good answer. The candidate has examined the two polymers carefully and deduced the three monomers correctly. However, the functional group at the left-hand end of the first monomer of M is not shown correctly. The $-CO_2H$ group in the first monomer is written as if it is at the right-hand end of the molecule. It should be written as HO_2C-.

(c) (i) $CH_2{=}CH_2$ ✗

This is incorrect and ignores the point in the question that the molecule has to have two functional groups.

(ii) Covalent bond ✗

This is not sufficient to score the mark — see Candidate B's answer.

(d) (i) $H_2NCH_2COHN\underset{\large CH_3}{\underset{|}{C}}HCO_2H$ ✓

This is the correct formula for the dipeptide gly–ala. However, it does not show the peptide bond in a displayed form, so the second mark is lost.

(ii) It is not optically active ✓.

This is correct.

(e) (i) Seven ✗

No. Seven amino acids are linked by six peptide bonds.

(ii) 575 ✓

This is correct.

(iii) Refluxed with acid ✓.

This is partly correct, for 1 mark. The process is refluxing with acid, but it matters which acid. The candidate should have mentioned concentrated hydrochloric acid.

Candidate B

(a) Attracted to water ✓.

This is correct.

(b) Monomer of L: $H_2C{=}CHOH$ ✓

Monomers of M: $HO_2CCH(OH)CH(OH)CO_2H$ ✓
and $H_2NCH_2CH(OH)CH_2NH_2$ ✓

This is a very good answer. All the monomers are identified correctly and their structures shown accurately.

(c) (i) $HO_2C{-}CH_2{-}CO_2H$ ✓

This is correct.

(ii) Covalent ester linkage ✓

This is correct.

(d) (i)

$$H_2NCH_2C(=O)\text{–}N(H)CH(CH_3)CO_2H$$ ✓

This good answer shows both the dipeptide gly–ala and the peptide bond in displayed form.

(ii) Glycine is not chiral ✓

This is correct.

(e) (i) Six ✓

This is another correct answer.

(ii) 565 ✗

This is an unfortunate miscalculation.

(iii) Reflux ✓ with concentrated hydrochloric acid ✓.

The answer is correct and complete.

Question 7

Chemical reactions in the body take place below 40 °C. They are able to do so because of the wide range of enzymes — biological catalysts — present in the body.

(a) What sort of macromolecules are enzymes? (1 mark)

(b) Fever, when the body temperature exceeds 40 °C, can be very dangerous because many enzyme reactions are no longer effective. Explain why this is the case. (2 marks)

(c) The diagram shows the reaction pathway of a reaction that is normally catalysed by an enzyme, but when the enzyme is *not* present. Sketch on the diagram the pathway when the enzyme is present.

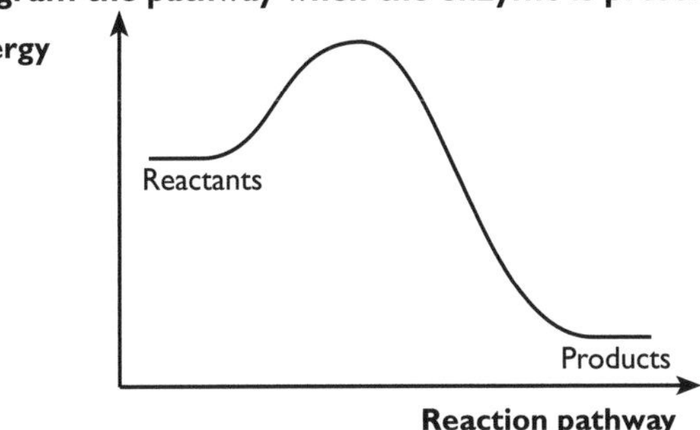

(2 marks)

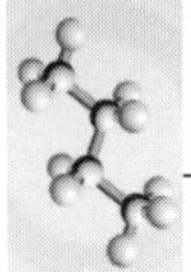

(d) The effectiveness of enzyme-catalysed reactions can be reduced by substances known as *inhibitors*. Explain the difference between a competitive and a non-competitive inhibitor. (2 marks)

(e) The graph below shows how the rate of an enzyme-catalysed reaction varies with substrate concentration at a given temperature.

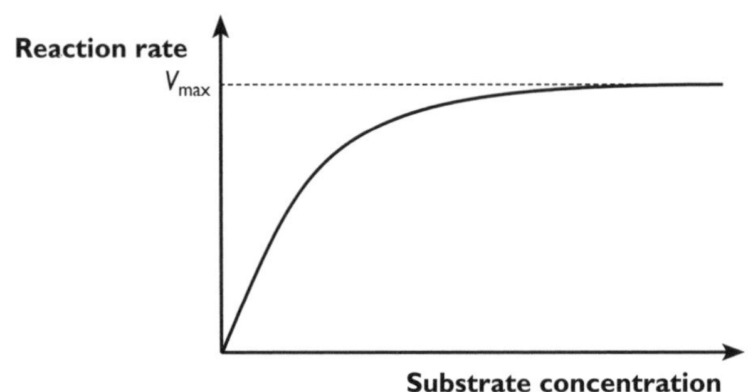

On the diagram sketch curves to show the effect on this reaction of:

(i) a competitive inhibitor, labelled C

(ii) a non-competitive inhibitor, labelled N. (4 marks)

Total: 11 marks

Candidate A

(a) Proteins ✓

Correct, enzymes are usually polypeptide or protein molecules.

(b) Proteins break down easily at low temperatures ✗ ✗.

No. A protein molecule does not break down at temperatures in the low 40s. Some of the weaker bonds present are disrupted, which changes the overall structure.

(c)

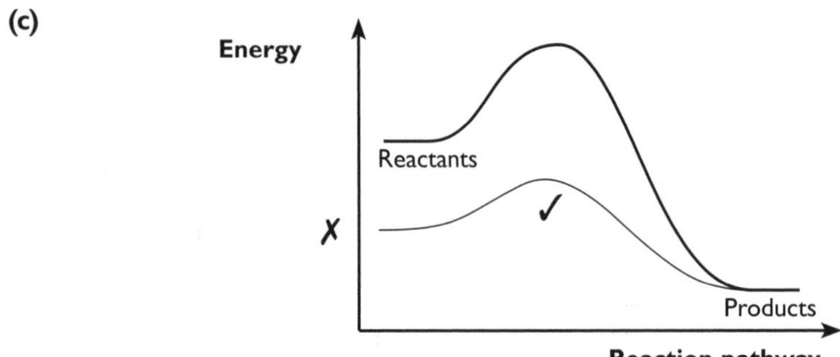

This answer is partly correct — the activation energy is shown correctly as lowered. However, the energy of the reactants should be the same as in the uncatalysed reaction.

(d) A competitive inhibitor binds temporarily to the active site ✗ but a non-competitive inhibitor binds permanently ✗.

Although the candidate clearly knows something about competitive and non-competitive inhibitors, the thinking is a little muddled. 'Competitive' means that the inhibitor is competing with the substrate for the active site. 'Non-competitive' means that there is no competition, so the effect must be different.

(e)

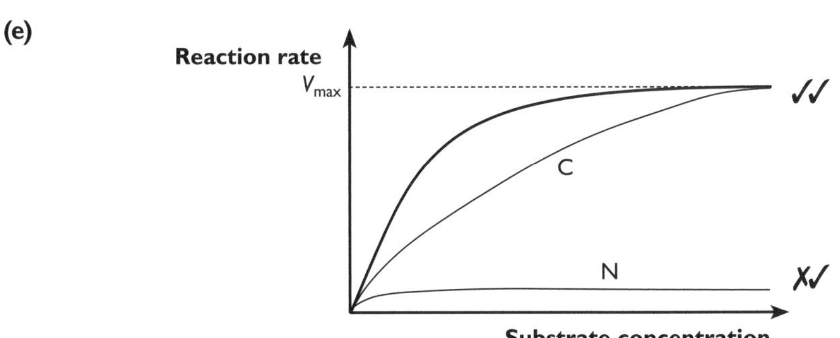

The graph for a competitive inhibitor is completely correct, i.e. with the same V_{max} and a different slope (K_m). However, the graph for a non-competitive inhibitor should show the same slope (K_m), but a different V_{max}. Only one of these conditions is met.

Candidate B

(a) Protein molecules ✓.

The answer is correct.

(b) At temperatures around 40 °C, the bonds forming the tertiary structure of proteins are disrupted ✓. This means that the 3-dimensional structure of the enzymes changes and the substrate no longer fits the active site ✓.

This is a very good answer. the reason for the change in activity of enzymes above 40 °C is explained well.

(c)

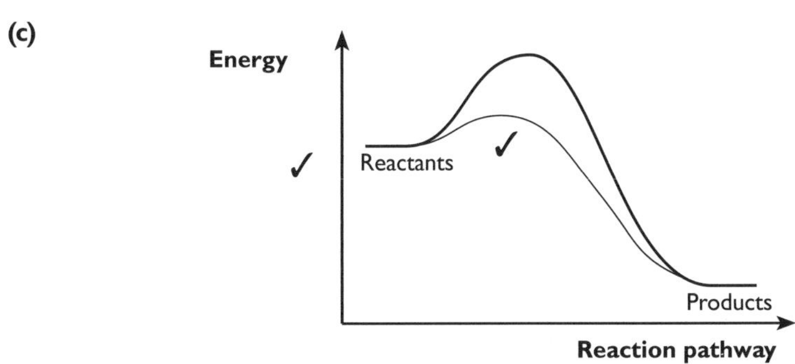

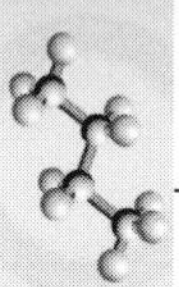

 This good sketch shows that although the activation energy is reduced by the presence of the enzyme, the energy of both the reactants and products is not changed.

(d) A competitive inhibitor competes with the substrate to bind to the active site ✓. A non-competitive inhibitor binds elsewhere in the molecule but changes the shape of the active site making it harder for the substrate to bind there ✓.

This very good answer highlights accurately the differences between competitive and non-competitive inhibitors.

(e)

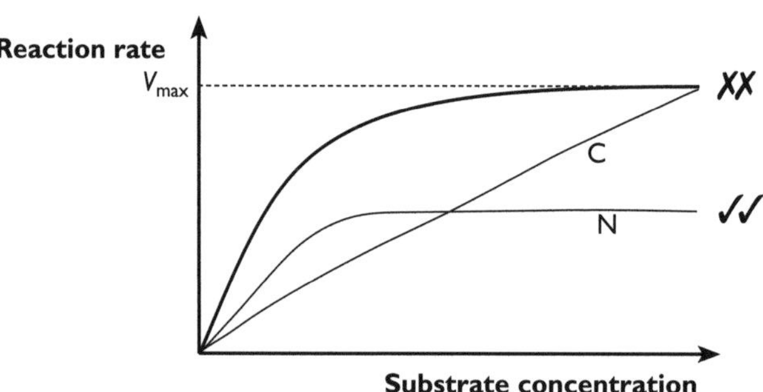

This is a pity. The candidate has remembered the effect of a non-competitive inhibitor, but has not thought through the line for the competitive inhibitor. It cannot be a straight line that stops suddenly.

Question 8

Modern instrumental techniques play an increasingly important role in determining the structures of organic compounds.

(a) An aromatic compound R has a mass spectrum in which the M and M + 1 peaks are in the ratio 10:0.9. The M peak is at *m/e* 122. Analysis of the compound gave the following composition by mass: C, 78.7%; H, 8.2%; O, 13.1%.

Showing your working:

(i) use the data to determine the empirical and molecular formulae of compound R (3 marks)

(ii) use the M and M + 1 data to confirm how many carbon atoms are present in compound R. (2 marks)

(b) The NMR spectrum of compound R is shown below.

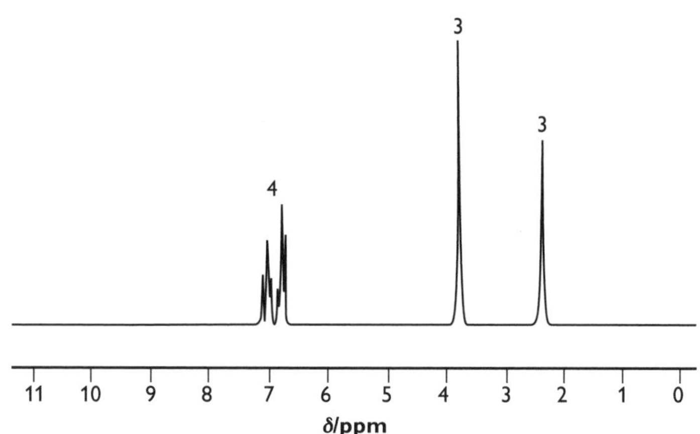

Use the *Data Booklet* to identify the types of proton present in compound R and hence deduce its structure. You should explain how you reach your conclusion. (4 marks)

(c) Compound S, an isomer of R, gave the NMR spectrum below.

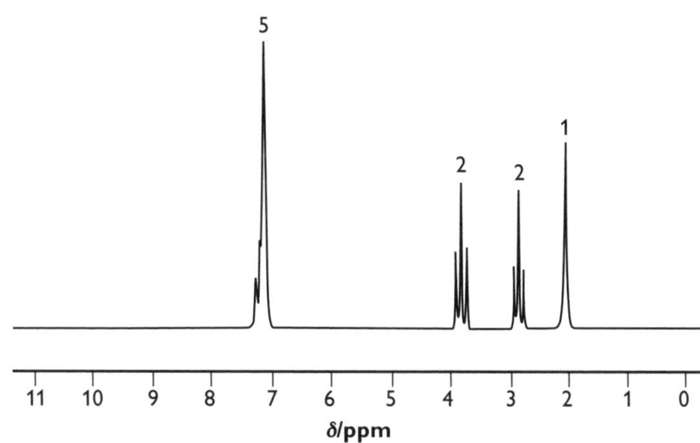

On adding D_2O to a sample of S and re-examining the NMR spectrum, the peak at $\delta = 2.0$ was found to have disappeared.

(i) Suggest a structure for compound S. (2 marks)

(ii) Explain why the peak at $\delta = 2.0$ disappears when D_2O is added. (1 mark)

Total: 12 marks

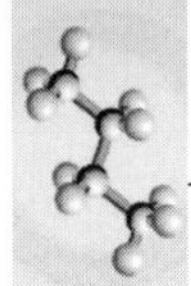

Candidate A

(a) (i)

Element	C	H	O
%	78.7	8.2	13.1
A_r	12	1	8 ✗
$\%/A_r$	6.6	8.2	1.6
Ratio	4 ✓	5	1

Based on the above, this compound would have an M_r of 69 and the empirical formula of C_4H_5O, but we know that R has a mass peak at 122. This suggests we need to double the ratio to give $C_8H_{10}O_2$. ✓ (ecf)

It is a pity that the candidate made the mistake of using the proton number rather than A_r for oxygen. This will have a knock-on effect in other parts because there should only be one oxygen in the formula.

(ii) If the heights of the M and M + 1 peaks are in the ratio 10:0.9, then the number of carbon atoms is

$$n = \frac{100 \times 0.9}{1.1 \times 10}$$ ✓ = 8.2 or 8 carbons ✓

The candidate has correctly used the M:(M + 1) ratio to calculate the number of carbons. This confirms the number obtained in the empirical formula calculation.

(b) δ = 2.4: single peak, 3 protons — methyl group attached to benzene ring ✓

δ = 3.9: single peak, 3 protons — methyl group attached to oxygen ✓

δ = 7.0: complex peaks, 4 protons — hydrogens joined to benzene ring ✓

The structure of compound R is:

CH_3—O—(benzene ring)—CH_2—OH ✗

The assignment of the δ values to correct proton types is correct, but now does not correspond to the structure shown for R. This is a result of the incorrect calculation of the number of oxygen atoms present.

(c) (i) The structure of S is:

HO—(benzene ring)—O—CH_2—CH_3 ✗✗

This structure is an isomer of R, but the arrangement of protons does not match the NMR spectrum of S, and no explanation is given.

(ii) This proton must be an –OH proton that exchanges with D, which does not show in an NMR spectrum ✓.

This is correct.

Candidate B

(a) (i)

Element	C	H	O
%	78.7	8.2	13.1
A_r	12	1	16
% / A_r	6.6	8.2	0.81
Ratio	8 ✓	10	1 ✓

The empirical formula is $C_8H_{10}O$, and since this has an M_r of 122, this must also be the molecular formula ✓.

This calculation has been carried out correctly. The empirical formula matches M_r and hence the empirical and molecular formulae are the same.

(ii) Number of carbon atoms (n) is

$n = \frac{100 \times 0.9}{1.1 \times 10}$ ✓ = 8.2 or 8 carbons ✓

The calculation of the number of carbon atoms from the M:(M + 1) peaks confirms that deduced in part (i).

(b) δ = 2.4: single peak, 3 protons — methyl group attached to benzene ring ✓

δ = 3.9: single peak, 3 protons — methyl group attached to oxygen ✓

δ = 6.9: complex peaks, 4 protons — hydrogens joined to benzene ring ✓

The structure of compound R is:

H_3C–(benzene ring)–O–CH_3 ✓

The assignment of δ values to correct hydrogen atoms means that the structure shown for R matches the NMR spectrum.

(c) (i)

(benzene ring)–CH_2CH_2OH ✓✓

δ = 2.0 — single proton attached to –OH (exchanges with D_2O)

δ = 2.9 — 1:2:1 triplet, so adjacent to $-CH_2-$ (similar to $-CH_3$ attached to benzene)

δ = 3.8 — 1:2:1 triplet, so adjacent to $-CH_2-$

δ = 7.2 — five identical protons attached to benzene ring

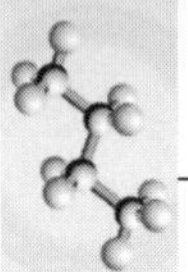

The structure shown for S is an isomer of R, and the explanation of the NMR spectrum is consistent with this structure.

(ii) This proton must be an –OH proton that exchanges with D, which does not show in an NMR spectrum ✓.

This is correct.

Question 9

A great deal of money is spent researching new drugs to combat diseases and medical conditions such as cancer. One of the challenges is to target drugs better so that they do not affect healthy cells, and also so that smaller doses can be given. Another challenge is to improve delivery of a drug to the site of the disease so that less drug is broken down on its way around the body.

(a) Drugs taken by mouth pass through the body's digestive system. In particular, the acid conditions in the stomach mean that drugs may be broken down there. The compound EPCG extracted from green tea shows promising anti-cancer properties when given by injection.

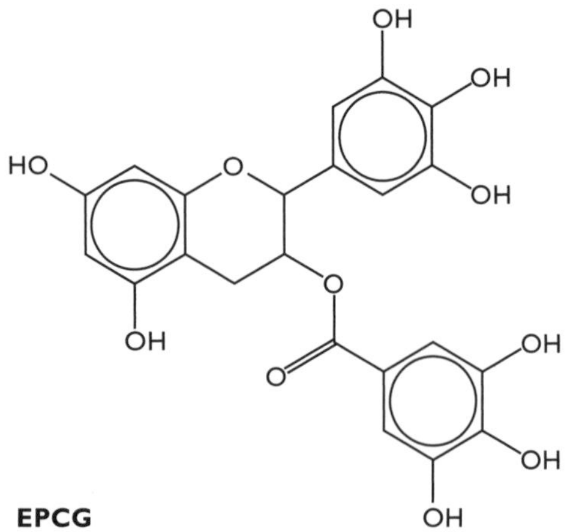

EPCG

(i) Circle *the functional group* that would be hydrolysed on passage through the stomach. (1 mark)

(ii) This compound is very soluble in aqueous solutions. Explain why this is the case. (2 marks)

(iii) Suggest another functional group that would be hydrolysed easily on passage through the stomach. (1 mark)

(b) Another way of ensuring that drugs reach their target is to protect them on their passage through the body. One way to do this is to

enclose the drugs in liposomes. These are tiny spheres made up of phospholipids, compounds which are naturally found in cell membranes.

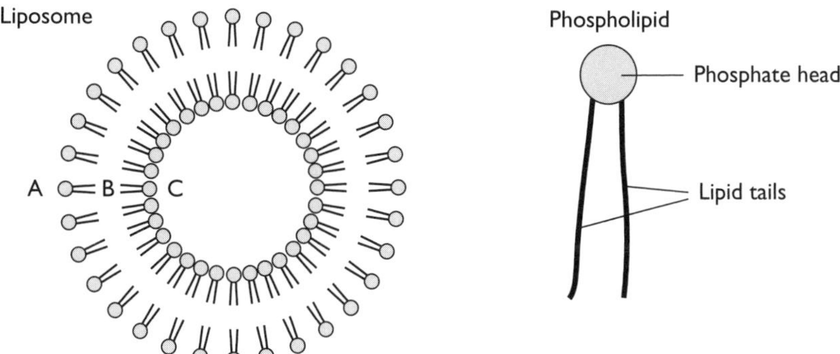

(i) Identify where on the liposome the drug EPCG would be carried, explaining your answer. (2 marks)

(ii) Suggest why there is so much interest in liposomes in the drug industry. (1 mark)

Total: 7 marks

Candidate A

(a) (i)

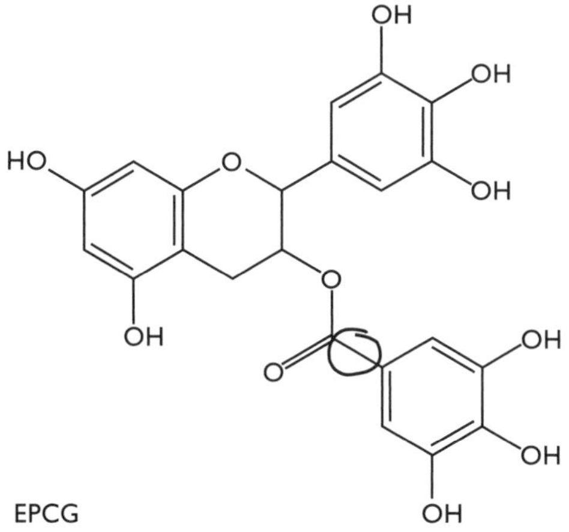

EPCG ✗

The candidate has not circled the ester linkage, which is the only group in EPCG that can be hydrolysed.

(ii) It contains lots of –OH groups ✓.

The statement is correct, for 1 mark. However, there is no explanation given. It is important to remember that 2 marks in a question means that two points are needed in the answer.

(iii) A peptide ✓

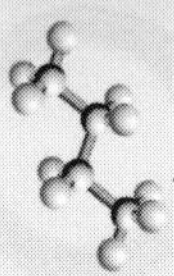

Although this is not strictly correct (peptide bonds are only usually referred to in protein chemistry) a peptide bond is effectively the same as an amide bond and thus the mark is awarded.

(b) (i) It would be carried at point **A** ✗ linked to the phosphate group.

Although the candidate has recognised that the compound would be linked to the phosphate group, at point A the drug would be exposed to reaction as it passed through the body.

(ii) They can carry drugs to where they are needed ✗.

This is too vague to gain a mark.

Candidate B

(a) (i)

EPCG ✓

The ester linkage is circled, which is correct.

(ii) It contains many –OH groups ✓ which can hydrogen bond with water ✓.

This is a correct statement about the structure and a correct reason for its solubility.

(iii) An amide ✓

This is correct.

(b) (i) It would be carried at point **C** linked to the phosphate group ✓. Here it would be protected from attack during its passage through the body ✓.

This good answer shows an understanding of the bonding and the reason for enclosing the drug in the liposome.

(ii) They can carry both hydrophilic and hydrophobic drugs ✓.

The candidate understands the nature of liposomes and the different environments that they contain — a good answer.

'Try this yourself' answers

(1) **(a)** $^{23}_{11}Na$ **(b)** $^{19}_{9}F$ **(c)** $^{32}_{16}S$ **(d)** $^{52}_{24}Cr$ **(e)** $^{39}_{19}K^{+}$

(2) **(a)** 0.5 mol **(b)** 1.5 mol **(c)** 0.25 mol

(3) **(a)** 4 g **(b)** 14 g **(c)** 7 g **(d)** 11 g

(4) **(a)** RbO **(b)** $C_3O_3H_8$

(5) 1.3 g

(6) 3.7 g

(7) 9.0 g

(8) 125 cm^3 of O_2; 75 cm^3 CO_2

(9) 1.59 g of PbO_2

(10) 1200 cm^3 of CO_2

(11) 0.110 mol dm^{-3} and 6.19 g dm^{-3}

(12) 0.15 mol dm^{-3}

(13) **A** $^{23}_{11}Na^{+}$; **B** $^{23}_{11}Na$; **C** $^{23}_{11}Na^{-}$; **D** $^{24}_{12}Mg$; **E** $^{25}_{12}Mg$; **F** $^{26}_{12}Mg$

(14) Group III; there is a large jump in energy after the third electron is removed.

(15)

Element	Charge on the ion	Electron configuration
Magnesium	+2	$1s^2 2s^2 2p^6$
Lithium	+1	$1s^2$
Oxygen	−2	$1s^2 2s^2 2p^6$
Aluminium	+3	$1s^2 2s^2 2p^6$
Fluorine	−1	$1s^2 2s^2 2p^6$
Sulfur	−2	$1s^2 2s^2 2p^6 3s^2 3p^6$

(16) **(a)** **(b)**

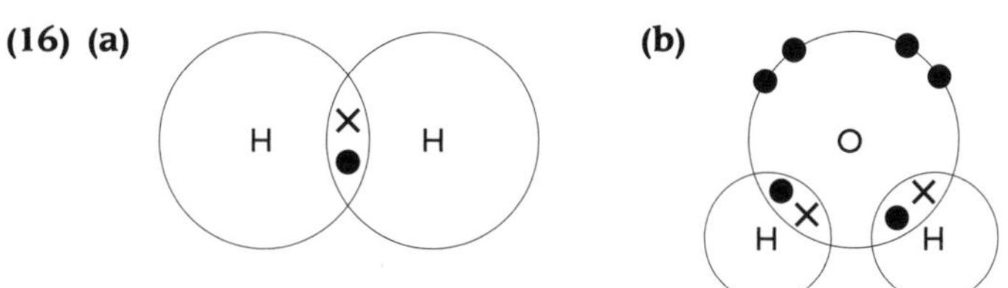

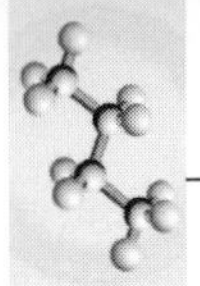

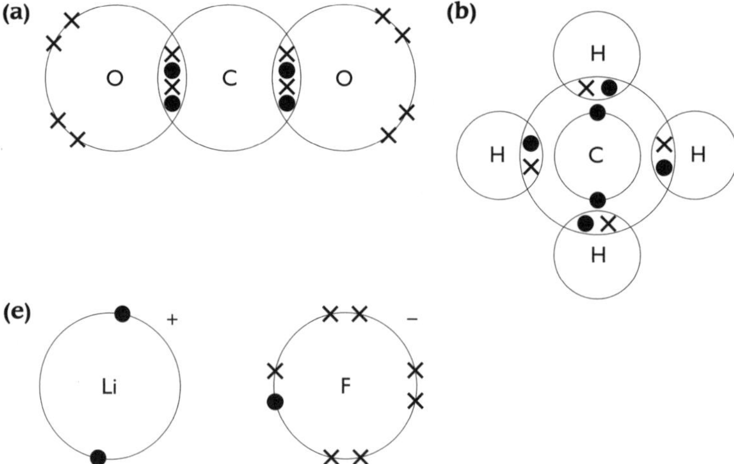

(17)

Material	Intermolecular force
Methanol, CH_3OH	Hydrogen bonding
Magnesium oxide, MgO	Ionic
Iodine chloride, IC*l*	Dipole–dipole/van der Waals
Argon, Ar	van der Waals
Aluminium, A*l*	Metallic

(18) It has a low melting point, so bonds between the particles in X are weak. It is an electrical insulator both as a solid and when molten, so it cannot be a metal or have ionic bonds. It dissolves in cyclohexane, which is covalently bonded. Therefore, the data suggest that X has covalent bonds within the molecules and probably van der Waals forces or dipoles between molecules. For example, paraffin wax.

(19) **(a)** An explosion, e.g. the ignition of hydrogen.

(b) Burning magnesium ribbon in air.

(c) Putting two different metals into an electrolyte, e.g. dilute sulfuric acid and connecting the metals with a wire.

(20) **(a)** +3 **(b)** 0 **(c)** +6

(21) $KMnO_4$ — Mn is +7; MnO_2 — Mn is +4; K_2MnO_4 — Mn is +6

(22) **(a)** +0.80 + (+0.76) = 1.56 V **(b)** −0.13 + (+2.38) = 2.25 V
(c) +0.80 + (−0.34) = 0.46 V

(23) **(a)** +2.38 + (+1.36) = 3.74 V **(b)** +0.13 + (+1.07) = 1.20 V
(c) +1.36 + (−0.77) = 0.59 V **(d)** +1.36 + (−1.07) = 0.29 V

(24) **(a)** **(i)** Since the reaction is exothermic, the equilibrium will move in the direction which absorbs energy, i.e. to the left.

(ii) Since the reaction has fewer molecules on the right than on the left, increasing the pressure will favour a shift to the right.

(b) (i) The reaction consumes CH_3CO_2H in moving left to right, so this is favoured if the acid concentration is increased.

(ii) Since the reaction produces water in moving from left to right, the removal of water will favour a shift in this direction.

(25) $H_2CO_3 + H_2O \rightarrow HCO_3^- + H_3O^+$

$$\underset{\text{Acid}}{H_2CO_3} + \underset{\text{Base}}{H_2O} \longrightarrow \underset{\text{Base}}{HCO_3^-} + \underset{\text{Acid}}{H_3O^+}$$

(26) J cannot be a gas because when it is in powdered form it reacts with oxygen. J forms a liquid chloride so it must be a non-metallic element. Only **silicon tetrachloride** hydrolyses to give an acidic solution and an insoluble solid.

$Si + O_2 \rightarrow SiO_2$

$Si + 2Cl_2 \rightarrow SiCl_4$

$SiCl_4 + 2H_2O \rightarrow SiO_2 + 4HCl$

(27) The graph shows a very large increase after the fifth ionisation energy, which suggests that the sixth electron must come from an inner shell. The removal of the first two electrons requires slightly less energy than the next three, which suggests that these are the two 4s electrons. Therefore, the element is vanadium.

(28) $MnO_4^- + 8H^+ + 5e^- \rightarrow Mn^{2+} + 4H_2O$

$O_2 + 2H^+ + 2e^- \rightarrow H_2O_2$

Multiply the first equation by 2: $2MnO_4^- + 16H^+ + 10e^- \rightarrow 2Mn^{2+} + 8H_2O$

Multiply the second equation by 5 and reverse it: $5H_2O_2 \rightarrow 5O_2 + 10H^+ + 10e^-$

Adding gives: $2MnO_4^- + 5H_2O_2 + 6H^+ \rightarrow 2Mn^{2+} + 5O_2 + 8H_2O$

(29) Assume that chromium is in the same oxidation state in each complex. Since the complexes are octahedral they must be of the type $[CrL_6]^{n+}$. Of the three ligands, OH^- gives the smallest d-orbital splitting, then H_2O and finally NH_3. The complex with hydroxide is green, the complex with water is blue and the complex with ammonia is purple.

(30) 2-hydroxypropanoic acid

(31) **(a)** 1-chlorobutane **(b)** 2-hydroxypropane or propan-2-ol
(c) Hexanoic acid **(d)** 3-methylbutylamine

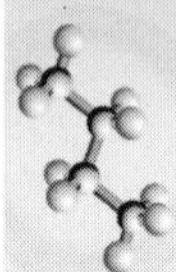

(32) (a) $H_3C—CH(Br)—CH_2—CH_3$ **(b)** $CH_3CH_2—C(=O)NH_2$

(c) $H_2C=O$ (H, H on C=O) **(d)** $H_3C—C(=O)—O—CH_3$

(33) (a) $CH_3CH_2CH_2CHO$ (or $CH_3CH_2COCH_3$)

(b) Examples showing a ketone (or aldehyde) and an enol — for example, $CH_3CH_2COCH_3$ and $CH_3CH{=}CHCH_2OH$

(c) Including butanal, there are 9 isomers excluding *cis-trans*, and 12 isomers including *cis-trans*

(34) (a) 10 **(b)** 3 pairs

(35) If aqueous bromine is used, both Br and OH can add to the double bond forming $CH_2Br–CH_2OH$.

(36) (a) Ethyl propanoate **(b)** Propyl methanoate
(c) Methyl butanoate **(d)** Propyl ethanoate

(37) K is an aldehyde because it reacts both with 2,4-DNPH and with Fehling's solution. It does not react with alkaline aqueous iodine so it cannot contain the $CH_3C{=}O$ group. Thus **K** must be $CH_3CH_2CH_2CHO$ and **J** could be butan-1-ol, $CH_3CH_2CH_2CH_2OH$, or 2-methylpropan-1-ol, $(CH_3)_2CHCH_2OH$.

(38) With phenylamine, the lone pair of electrons on the nitrogen atom become delocalised with the π-electrons on the benzene ring making it less easy to protonate. In ethylamine, the ethyl group pushes electrons towards the nitrogen enhancing its ability to be protonated.

(39) In condensation polymerisation:
- a small molecule is eliminated
- the monomer(s) need two different functional groups

In addition polymerisation:
- the empirical formula of the polymer is the same as that of the monomer
- an unsaturated carbon–carbon bond must be present

(40) (a) Asp **(b)** Pro **(c)** Lys

(41) Any one of the following triplets:

(a) AGU AGC UCU UCC UCA UCG **(b)** UGU UGC **(b)** UAU UAC